A HANDBOOK OF
COMMUNITY
GARDENING

A HANDBOOK OF
COMMUNITY GARDENING

—◆—

BY BOSTON URBAN GARDENERS

EDITED BY SUSAN NAIMARK

Charles Scribner's Sons ◇ *New York*

We dedicate this book to the community gardeners,
who have taught us so much,
in the hope that it will share at least
some of what we have learned.

Unless otherwise indicated,
all drawings are by Susan Naimark.

Copyright © 1982 Boston Urban Gardeners, Inc.

Library of Congress Cataloging in Publication Data

Main entry under title:

A handbook of community gardening.

Bibliography: p.
Includes index.
1. Community gardens. 2. Gardening. I. Naimark,
Susan. II. Boston Urban Gardeners. III. Title:
Community gardening.
SB457.3.H26   635   81-23302
ISBN 0-684-17466-9   AACR2

This book published simultaneously
in the United States of America and Canada.
Copyright under the Berne Convention.

Printed in the United States of America.

1 3 5 7 9 11 13 15 17 19   F/C   20 18 16 14 12 10 8 6 4 2

The quotation from Robert Frost's "Mending Wall"
in Chapter 11 is from The Poetry of Robert Frost,
edited by Edward Connery Lathem. Copyright 1930,
1939, © 1969 by Holt, Rinehart and Winston.
Copyright © 1958 by Robert Frost. Copyright ©
1967 by Lesley Frost Ballantine. Reprinted by permission
of Holt, Rinehart and Winston.

Acknowledgments

———— ◆•◆ ————

For an organization to write a book is no easy task. Although the individual chapters in *A Handbook of Community Gardening* represent "the best of BUG," many unnamed members of Boston Urban Gardeners have contributed to our collective community gardening wisdom with their own experiences and skills.

I'd like to thank those who have been quoted throughout the book, not just for their words that have been put into print but for their dedication to community gardening: Ray Almeida, Ed Cooper (BUG's president), Esther Diggs, Bud Dupont, Isaac Graves, Lloyd A. Harding, Gareth Kincaid, Tom Libby, and Mrs. Francis Watkins. For assistance in gathering these insights I'd like to thank Gail Johnson and Charlotte Kahn.

For making the book possible in the first place, thanks go to Dan Burnstein (BUG's lawyer) and the Boston Urban Gardeners Board of Directors.

For assisting with information used in the book, thanks to Shelley Abend, the Boston Natural Areas Fund, David Lovler, Julie Stone. Participants in BUG's Wednesday coalition meetings contributed support and feedback.

For editorial assistance thanks go to Martha Martin and, most important, to Judith Wagner, who patiently reviewed each draft and added her own insights, which come from her years of leadership in the community gardening movement.

Susan Naimark
Editor

Notes on Contributors

Susan Naimark ("The Plants," "Solar Devices to Extend the Growing Season," "A Community Canning Center," "A Food Co-op Community Garden," coauthor of "Fencing") is director of the Southwest Corridor Community Farm. The farm, organized in 1977 by neighborhood residents, is a community garden and educational center hosting demonstrations on gardening, composting, and soil building practices. Its solar greenhouse provides seedlings to the entire Boston gardening community, which helps support the farm's work. The editor and illustrator of this book, Susan does free lance work in graphic design and illustration.

Nina Gomez-Ibanez ("Growing with Kids") was a founder of the large community garden in Charlestown, formally incorporated in 1980 as Gardens for Charlestown, Inc. Each year the group hosts a garden tour of neighborhood and backyard gardens. Nina is also an active member of the BUG Board of Directors, concentrating on garden productivity and beautification.

Patricia Grady ("Land Ownership Alternatives," coauthor of "Fencing") is a founder of the Mission Hill Community Gardens and is a catalyst in organizing the Mission Hill Community Land Trust. Patricia is currently director of a Community Food and Nutrition Program at Boston's antipoverty agency.

Charlotte Kahn ("Historical Roots: The History of Community Gardening," coauthor of "Locating Land" and "Site Layout") is a founder of the South End

Garden Project, the Southwest Corridor Community Farm and Boston Urban Gardeners, and codirected BUG for four years. Currently BUG's director, Charlotte has concentrated on developing an effective coalition of urban garden groups in Boston and on researching historical and political antecedents for the urban open space policies BUG supports today.

Pat Libby ("Experiences in Community Gardening") is the assistant director of BUG. Former director of the Food and Nutrition Program at Boston's antipoverty agency and an active board member, Pat has concentrated her energies on community-based food action projects. She has also served for several years as BUG's representative to the Massachusetts Food and Agriculture Coalition (Mass FAC).

Jane Lueders ("Landscape Planting") is a professional landscape architect who volunteered her services to help several community gardening groups improve their gardens' designs.

Martha Martin ("Keeping the Garden Going," coauthor of "Vandals in the Garden") has been active in both Boston Urban Gardeners, as a staff and board member, and in Dorchester Gardenlands Preserve and Community Development Corporation. Martha has directed her efforts toward direct marketing and land trust activities.

Jean E. Morse ("Water," coauthor of "Site Layout") is one of the prime movers behind the organization of the Mission Hill Community Gardens. The group was formally incorporated in 1981 as a Community Land Trust and is now seeking to purchase two garden sites.

Susan Redlich ("The Public Role," "Problem Solving through Coalition Building") has been the director of the Massachusetts State Department of Agriculture's Division of Land Use since its inception in 1977 and has been active in BUG since it began as well. The Division of Land Use has been instrumental in drafting and securing passage of legislation to secure farmland, increase the state's food self-reliance, and create direct marketing opportunities for farmers. The Division has worked closely with BUG on the development of a large-scale compost facility in Boston and on myriad urban agricultural issues.

David Rosenmiller ("Permaculture in Community Gardening") is a student of environmental design at the College of the Atlantic who has worked with BUG both as a volunteer and as a staff member on various gardening projects, including developing a composting facility and researching the design of an edible urban landscape.

Linda Roth ("Soil," "Compost") is a soil scientist and currently works for the Suffolk County Extension Service. The Extension Service provides a variety of technical and educational services related to gardening, including a free soil testing service for county residents, written resource materials, and a horticultural hotline. Linda is taking a leave of absence from the Extension Service to teach soil science in Nicaragua.

Virginia Scharfenberg (coauthor of "Locating Land") is President of Dorchester Gardenlands Preserve and Community Development Corporation, and has concentrated on establishing permanent gardens in a community land trust and on linking urban food consumers to rural food producers through farmers' markets and direct marketing.

Judith Wagner ("Community Gardens: Growing Together," "Basic Steps of Organizing a Community Garden Group," coauthor of "Vandals in the Garden") is a founder of Boston Urban Gardeners and was codirector of BUG for four years, and has documented the economic development potential of urban agriculture in a recent Masters of City Planning thesis. Currently a member of the BUG Board of Directors, Judith continues to be involved with community economic development and agricultural issues.

Greg Watson ("Resources: What You Need and Where to Find It," "Farmer's Markets") is an artist, environmentalist and educator, and has worked for the Massachusetts Department of Agriculture's Land Use Division and the New Alchemy Institute. Greg has been responsible for organizing many of Boston's Farmers' Markets and appropriate technology demonstration projects.

Contents

———◆•◆———

PART I   **An Introduction to Community Gardening**   1

  1. Community Gardens: Growing Together   3
     Judith Wagner

  2. Historical Roots: The History of Community
    Gardening   11
     Charlotte Kahn

PART II   **From Idea to Reality**   17

  3. Basic Steps of Organizing a Community Garden
    Group   19
     Judith Wagner

  4. Resources: What You Need and Where to Find It   26
     Greg Watson

  5. The Public Role   32
     Susan Redlich

PART III   **Site Selection and Development**   43

  6. Locating Land   45
     Charlotte Kahn and Virginia Scharfenberg

7. Site Layout   53
    Charlotte Kahn and Jean E. Morse

8. Soil   59
    Linda Roth

9. Compost   69
    Linda Roth

10. Water   74
    Jean E. Morse

11. Fencing   81
    Patricia Grady and Susan Naimark

12. Landscape Planting   88
    Jane Lueders

PART IV   Tending the Garden   95

13. The Plants   97
    Susan Naimark

14. Keeping the Garden Going   108
    Martha Martin

15. Vandals in the Garden   113
    Martha Martin and Judith Wagner

16. Growing with Kids   117
    Nina Gomez-Ibanez

17. Solar Devices to Extend the Growing Season   121
    Susan Naimark

**PART V   Beyond the Garden: Developing Local Food
            Systems**   127

18. Land Ownership Alternatives   129
    Patricia Grady

19. Problem Solving Through Coalition Building   136
    Susan Redlich

20. Farmers' Markets   140
    Greg Watson

21. A Community Canning Center   144
     Susan Naimark
22. A Food Co-op Community Garden   147
     Susan Naimark
23. Permaculture in Community Gardening   150
     David Rosenmiller
24. Harvesting Our Experiences   158
     Pat Libby

Tables   163
Notes   173
Annotated Bibliography   175
Index   177

Part I

---◆---

AN INTRODUCTION TO COMMUNITY GARDENING

Photo by Read D. Brugger

1

Community Gardens: Growing Together

Judith Wagner

A community garden is like a bumblebee. The laws of aerodynamics say that the way it's built, a bumblebee can't fly. That's why we use the bumblebee as our garden's symbol. Everyone told us we couldn't make a community garden work in the city, but we did!

—Gareth Kincaid,
Shangri-La Community Garden, Mattapan

Community gardening, an old practice that is enjoying a revival, is cropping up in unlikely places across the country. Like the spring's first surprising crocus, gardens are appearing where they might least be expected. In rural areas, neighbors come together to put an idle piece of farm or municipal land to use for family gardens. In suburban areas, people tend plots on the grounds of large companies or join groups using town land or even public parks for gardening. And in cities, where the very idea of a garden might seem to be as unlikely as a mirage, gardens are appearing on once-vacant lots, persisting under the most difficult conditions, like those courageous city plants that manage to push their way through asphalt and cement, using even the smallest cracks to gain a foothold.

Nationally, there is a measurable resurgence in the practice of home gardening in response to the pressures of the energy crisis and the accompanying drastic increase in the cost of basic necessities such as fuel, housing, and food. Gardening

3

is a uniquely beneficial pastime, providing not only the economic benefits of food production but, for many, a major source of recreation and outdoor enjoyment. Yet many people—whether they live in rural towns, suburban developments, or the heart of a city—do not have access to the most basic ingredient for gardening: land. It is these people who have revived the creative and practical idea of community gardening.

This book is about community gardening. It is a collection of information, ideas, and suggestions on why, where, when, and—most important—how to start a community garden. *A Handbook of Community Gardening* was written as a group effort by members of Boston Urban Gardeners, a coalition of people involved in urban community gardening in more than a dozen neighborhoods in Boston and in several surrounding towns. The experience we offer is from Boston, but we believe the information, examples, and experience we have collected will be useful to gardeners everywhere—even if they garden only in their back yard.

Community gardening is by definition a different experience from ordinary gardening because basic resources—land, water, even sunlight—must be shared. For that reason, this book is written to help community gardeners meet the challenges of group gardening. Nevertheless, many of the needs of community gardeners, including low-cost materials, do-it-yourself designs for basic systems, and low-maintenance and hassle-free methods, apply to all gardeners. The materials assembled in this handbook should be valuable to any gardener who wants resourceful ideas for developing a garden site.

"What Is a Community Garden?"

A community garden is: a place to go . . .
To learn and educate;
To feel as part of the natural earth (the bricks and mortar are the elements
  of the environment that create anxiety and pessimism);
To develop a mechanism for communication, information sharing and
  networking;
To preserve and provide for the highest and best use of valuable resources;
To reduce the expenditure of cash for food and exchangeable materials;
To have peace of mind.
A community garden is a place to go to communicate with earth, nature and
  oneself.

—Lloyd A. Harding, organizer of the
Forest Street Garden, Roxbury

A community garden is first and foremost a garden where people share basic resources—land, water, and (especially in the city or densely built areas) sunlight. Although community gardens vary enormously as far as the extent of group activity, this basic sharing of resources makes some degree of cooperation necessary.

In Boston, community gardens have multiplied at an amazing rate. In the past four years, the number of community gardens has grown from 20 to more than 120, together covering over fifty acres of land. The variety of gardens ranges from the tiny garden in the heart of Boston's Chinatown (the city's most densely populated area, where no other land is available) to the expansive Fenway gardens, which provide space for more than five hundred gardeners in one of Boston's oldest public parks. Although the gardens are as individual as the people who tend them, community gardens share certain characteristics. To Mrs. Francis Watkins of the Lower Roxbury neighborhood, a community garden means

> A mixing with people, learning with each other. A garden cuts down on the cost of living, you can grow flowers (everybody should have flowers!) and vegetables. It gives everyone something to talk about. Each one knows a little that you don't know. You can tell each other what you know.

Gardens are a place for sharing knowledge and experience as well as natural resources.

For Tom Libby, who is struggling along with his neighbors to improve his section of Dorchester (a neighborhood hard hit by disinvestment and arson, which have left more than 100 vacant parcels in a twenty-two–street area), a community garden means

> No one person owns it. It's the cooperative aspect of the thing. It's not only good for the people who use it, but the fact that it's on a vacant lot improves the whole community. It combines in the best way the selfish interests of neighbors—exercise, food, recreation—with the more altruistic reasons for a garden: the land looks better for everybody and encourages the whole neighborhood.

A community garden, then, meets many needs and serves more than the one purpose of providing residents with a place to grow food. The community garden is a common ground for people who share a love of the earth and a concern for and an understanding of the life cycles and intricacies of natural systems. The experience of community gardening transcends barriers of language, age, sex, race, ethnic origin, and the many other cultural divisions that come between us.

In the gardens in Boston, the gardeners range from lawyers and bus drivers to factory and clerical workers, carpenters, and homemakers. The combined skills of these people often make it unnecessary to look beyond the gardening group for solutions to practical problems.

People attracted to community gardens are prepared to share not only their varied skills and experience but other common characteristics as well: a remarkable resourcefulness, a do-it-yourself approach, a basic optimism, a respect for natural systems and other people, the ability to make do with whatever materials are at hand. These traits make gardeners especially adept at getting the most out of scarce resources, particularly in urban areas where the basic materials for healthy gardens (and people) are often in short supply.

Community gardens are the sites of a unique combination of activities: food production, recreation, social and cultural exchange, and the development of open space, community spirit, skills, and competence.

WHY COMMUNITY GARDENS?

Of the many reasons why people participate in community gardens, the savings on food costs is probably primary. Recent surveys show that the financial benefits of gardening rank first on a list of reasons for planting home and community gardens. Although there is widespread skepticism about the amount of food that can be raised on a community garden plot, particularly in the city where plots tend to be small, mounting evidence shows that a family can save a significant amount of money on its food costs by eating vegetables grown in even cramped conditions. The national average for savings is $250 per 600 square feet of garden;[1] in Boston, however, we have data showing that many gardeners save two or three times that amount. Using the most conservative figures as an example, Boston's community gardens (totaling approximately fifty acres of land) are producing over $235,000 worth of food. If all the gardens could achieve the higher levels of production that we have documented in the plots of especially skilled gardeners, that same number of gardens could produce over $1 million in food value.

If you find that public officials and others whose cooperation you need to start or support a community garden are among the skeptics regarding the financial benefits of gardening, the above facts and figures may help. Community gardening is, in its own quiet way, a notable economic force. For the many families across the country who do not own their own land or home, community gardening may be the only way to gain access to land for food production.

Why should people attempt to grow a portion of their food supply themselves? Isn't America the "breadbasket" of the world? Why should people bother to spend their time and energy in a garden plot?

These questions do not have simple answers. People garden for many reasons. But the economic reasons for gardening are directly related to complicated trends in our national economy. For one thing, food shows the largest increase in price of any basic necessity in the American economy except fuel. The average American family spends 25 to 31 percent of its household budget on food,[2] a proportion that increases as the family's income goes down. The increase in food prices can be traced to two major factors, both directly related to the energy crisis first felt dramatically in the United States in 1975. Over the preceding decade, much of our food production had been converted to large-scale agribusiness, which requires extensive mechanization and ever-increasing quantities of synthetic fertilizers. Both the machinery and the fertilizers are petroleum-dependent, and the cost of this basic ingredient has increased steadily with the price of oil. A second factor in the rise in food prices is the concentration of food production in a few areas of the country that have especially attractive conditions for agriculture on a large scale. Agricultural production in other areas of the country has been reduced to a few specialty products (maple syrup and apples in New England, potatoes in Idaho, butter in Wisconsin) or practically eliminated. Although this trend has had some advantages, it is dependent upon cheap energy for production and long-distance transportation to move food from the point of production to markets across the country. With the rising price of fuel, transportation, too, has become increasingly costly.

The long-distance marketing of food has resulted in other price-increasing trends such as the expensive packaging and processing required to keep food from spoiling in transit. In addition to these factors, the general rate of inflation, which affects everything from wages to raw materials to the cost of advertising, has kept food prices high. Together, all these increases are forcing food prices to levels that are pinching family budgets. People are finding that gardening helps ease the financial strain.

Happily, gardening provides a number of related benefits that for many people make the effort worthwhile for reasons beyond the economic advantages. One benefit is the sheer wonder and joy people find in helping things grow. For many of us there is nothing quite like the experience of putting a wrinkled, unlikely looking seed into the soil and watching it transform itself into a plant that bears food or flowers. Gardening is hard work, but it does involve nurturing and careful tending.

Gardening is also outdoor work, bringing people in contact with the sun, the air, the rain, and the earth. It can be recreational in the original sense of the word: people reestablish themselves as they become involved. Being in touch with the natural processes can change a person's perspective on day-to-day events. Instead of being annoyed that the forecast is for rain, making it necessary to carry a rain-coat or an umbrella, a gardener's first thought may be for the garden: "How we need that rain!"

In a community garden, the pleasure can extend beyond the strictly personal experience. There will often be at least one other person around when you tend your plot. Gardeners strike up friendships with a variety of people who share the enjoyment of growing plants. The garden can become a focal point for community events—picnics, barbeques, block parties, fairs, or bake sales. One large garden in Boston starts each spring with a May Day "Wake Up the Earth" festival, a time for neighbors to come together and celebrate the renewal of spring greenery.

Agriculture means more than cultivating the earth. It means passing on impor-tant strands of the social culture of the community. Ray Almeida, a community gardener of Cape Verdian descent who was born and raised and now gardens in Boston's South End, has this to say about the culture of community gardens:

> The Chinese plot out their land very differently than Latinos and Cape Verdians do. I'm talking about different cultivating techniques, but that in itself is not the principal thing. It's the way in which people learn, in dealing with those gardens, to wrestle with each other, with the way in which people begin to get at some basic understand-ings of what kinds of issues are at stake, and hold onto and develop and maintain some sense of control over this piece of turf which is our communal space, our Boston Common, that somehow or other has persisted in spite of all the onslaught of urban renewal and everything else.

Along with these fairly visible and direct benefits, community gardens make a number of other positive contributions, one of the most important being the improvement of environmental quality. Pollution and general degradation of the environment is a national problem. Even farmlands suffer serious deterioration due to the use of chemical and mechanical agricultural practices that increase soil erosion and the leaching, or carrying off by rain, of soil nutrients. Suburbs too often suffer from unplanned or unguided development, which interferes with nat-ural water run-off and other natural systems. In our cities the environment has been abused for years by dense building—resulting in the distortion of wind, rain-fall, and sunlight and creating unnatural heat and noise conditions—and by pol-lution from myriad sources.

Community gardens alone cannot solve these problems. But increased contact with the soil and natural systems does make people more aware of the fragile nature of the environment and increases their interest in protecting it wherever their gardens are. Gardens, especially in barren city environments, can make a significant contribution to the health of the city as a whole. It has been estimated that a city the size of Boston (roughly 625,000 people) needs at least five hundred acres of green space to replenish the amount of oxygen consumed by that population. Most of the city's open spaces are either parkland, which requires maintenance crews and expenses that strained city budgets cannot support these days, or recreational sites such as basketball courts, which have no greenery and so do not contribute at all to improving air quality. Gardens offer several advantages over both of these uses of open space. Gardens are managed and maintained by gardeners on a volunteer basis and do not require attention by municipal parks departments. The soil of gardens is well worked; it serves as a sponge for rain water, unlike some green space, which may be less absorbent as a result of hard-packed soil, sloping, or other conditions. Finally, the concentrated greenery in gardens restores oxygen to tired city air.

Some less visible advantages of community gardens are more social than environmental. Gardens provide a meeting ground for people of all sorts; they attract a certain type of hardy folk to a common project and encourage people to work together in other ways, providing them with a specific goal and visible proof of success when it is reached. These characteristics of community gardens increase the feeling of neighborhood community. "Stable" neighborhoods are often measured only in financial terms, but much of the "instability" of urban communities, in particular, is related to the long-term frustration of efforts to improve the community situation. The isolation of neighbors is both a cause and an effect of these failures. The community garden consistently bridges these gaps between neighbors. Efforts to make the neighborhood a true community now meet with modest success. The proliferation of garden greenery in cities speaks powerfully to people weary of cement, asphalt, steel, and abandoned lots. Community gardens show that people care, that they dare to invest in their community. They combine a refreshing environmental interest with an implicit political statement of commitment to land and community.

In a final word, community gardening is part of a serious struggle, the struggle to redistribute basic resources to people who will use them wisely and with respect for the general good. Community gardening is a small but serious challenge to many current policies and practices. It challenges the economically and ecologically destructive policies of agribusiness and local politics, which put profit before human needs—greenery, open space, fresh food. Community gardening challenges the social and economic structures that keep a vast number of urban and rural people from owning land and from gaining a small measure of control over their own lives. As Ray Almeida puts it:

> Somehow we have one foot firmly planted on this space, and if there is to be a last stand, if there is to be a last stand in so many of the urban neighborhoods in Boston, that poor folks are going to be involved in, then it's going to be around the issue of those gardens that they've invested their sweat in, that they've taught their children in, that they've taken food from, and where they have somehow stayed in touch with very basic, elementary, nonromantic, fundamental kinds of things about what it is to be a man, a woman, a head of house, a conveyor of the culture in a technological society that doesn't really value that all that much.

GROWING A COMMUNITY GARDEN

If you are interested in community gardens for one or more of the reasons mentioned above, the question now is how to get a garden started and, once started, how to keep it thriving. Because community gardening is a relatively new (or newly revived) activity in many communities, you may encounter many different problems, from financial and organizational to political. We enumerate these not to discourage you but to offer some common sense and experience on how to deal with these difficulties. We know it can be done, and we have collected the best advice we can offer to help you do it.

The first part of *A Handbook of Community Gardening* describes the initial organizing you will need to go through, from getting people together to locating the necessary resources.

Part II offers ideas and information on developing the garden site you have. In each case we have tried to be specific, aiming our suggestions toward low-cost and do-it-yourself methods.

Part III describes long-term needs for a garden and some of the important developments in it that you may want to encourage.

Finally, we offer glimpses of a vision for a full-scale food system that can grow from the community gardens in your area.

The entire BUG coalition, which has brought together its group experiences for this book, wishes you good growing and a fine harvest.

2

Historical Roots:
The History
of Community Gardening

Charlotte Kahn

In the frenetic beginnings of community gardening in Boston in the 1970s, there was little time for reflection. Our daily tasks involved moving mountains of earth from construction sites and dams in the suburbs to vacant lots in the inner city. In the rush to find, build, or invent fencing and water systems, we scarcely looked up, let alone across, history or oceans to find antecedents. Yet without much experience we knew what to do. Gardeners who walked out their front doors began to sift stones from the soil and divide the ground into family-size plots without even a discussion.

Half a decade later, when community gardens are commonplace in inner cities, suburbs, and small towns throughout the United States, we can take time to examine their historical precedents.

Community gardening has its deepest roots in the oldest and most universal form of human settlement: the tribe living in a self-sufficient village. With the exception of nomadic peoples, every civilization began with, and many still today rely for survival on, community-based subsistence agriculture. Villagers would sow and reap together, staying near the settlement for safety, growing enough to feed their families and, with luck, enough to lay some food aside for particularly hard times. In Asia, Africa, Europe and pre-Columbian America, the beginnings of agriculture were the same.

As civilizations became more urbanized, feudal and, later, commercial hierarchies took control of the land, exacting stiff fees and rents from tenant farmers in return for protection. Formerly communal or adjacent subsistence gardens gave way to large cultivated fields of grains, staples, and cash crops. Following the Renaissance in Europe, the spread of exploration, commerce, and money-based economies throughout the world further broke down ancient agricultural patterns. In Africa, rural people were ripped from their fields and villages and sold into slavery. In Europe, tenant farmers were forced off lands to which they had held common right for centuries to make way for enclosed, intensively cultivated commercial farms and pastures. Catering to the new economy, centralized farms were now more profitable than negotiated agreements with tenant farmers. Many rural areas were almost depopulated as landless people made their way to expanding urban centers.

The Industrial Revolution, which began in the mid-eighteenth century, accelerated this transition from self-sufficient agricultural to industrial, money-based economies. Community gardening became a way to retain ties to the land in an urbanized world. As rural people moved into the cities of Europe, some of them recreated an urban version of older feudal systems by renting plots of land on the outskirts of development. In England, plots were rented for one guinea. A record of one such "guinea garden" in Birmingham, England, from the mid-eighteenth century reveals a sight common to urban community gardeners today:

> From the west end of this area (north of the town centre) we enjoy a pleasing and lively summer-view over a considerable tract of land laid out in small gardens. This mode of applying plots of ground, in the immediate vicinity of the town, is highly beneficial to the inhabitants. They promote healthful exercise and rational enjoyment among families of the artisans, and, with good management, produce an ample supply of those whole-some vegetable stores, which are comparatively seldom tasted by the middling classes when they have to be purchased.[1]

As the Industrial Revolution progressed in Europe and the United States, gardened and farmed land around most cities was subdivided into residential, industrial, and commercial property. By the end of the nineteenth century, in the wake of overcrowding and severe epidemics, parks and gardens came to be seen as a public health necessity. In England, the Allotment Acts of 1887 and 1890 and the Local Government Act of 1894 required sanitary authorities in urban boroughs to provide space for community or "allotment" gardens. The Small Holdings and Allotments Acts of 1907 and 1908 gave rural residents access to garden plots in 500-square-yard parcels as "small holdings" or "allotments." The Agricultural Organization Society promoted this kind of agriculture and smoothed communication among farmers, small-holders, and allotment-holders with the government's support. Later, the standard rural allotment was reduced to 300 square yards.[2]

In the United States, a similar urban movement for public access—first to open space in general and then specifically to gardening space—grew in the late nineteenth century. Expanding metropolitan areas across the country set aside large tracts of land as permanent parklands. Despite increased use of urban open space for recreational purposes, cultivated areas near cities disappeared. Food was imported from distant market gardens to the central cities. As a result of the Panic of 1893, which created severe food shortages, Detroit initiated a unique form of unemployment relief by setting aside vacant city land for community gardens in which people were encouraged to grow staple crops. By 1895, 455 acres were under cultivation as "potato patches," and the city's initial $5,000 investment had produced $28,000 worth of produce.[3]

Around the turn of the century, waves of immigrants from rural homelands poured into the nation's cities. American cities were cosmopolitan, crowded, and expanding as never before. Agricultural land was again swallowed up, this time by electrified streetcar, sewer, water, and gas lines, which stretched across the countryside. Suburban sprawl began in earnest.

As urbanization and industrialization created new need and demand for public access to open space, horticultural societies and civic groups in Boston, Philadelphia, Cleveland, Minneapolis, and other cities created the School Gardens movement. As the Massachusetts Board of Education stated in its 1906 report:

> Not even in New York itself can territory be found more completely built over than our North and West Ends. Yet we have our gardens there. In 1902 the Schoolhouse Commission bought a lot of land adjoining the yard at the Hancock School in the North End, tore down the old tenement which stood thereon, and turned it over, bricks, mortar, and hole in the ground, to our committee. After the rubbish was removed we appealed to another city department for help. We asked the street department for twenty-five loads of sweepings, and from these (although after our good loam had gone on, boots, tin cans, and crockery appeared as a result of cultivation) we got a good deal of fertilizer gratis.
>
> In the West End the Board of Park Commissioners . . . granted the use of two strips of ground in the Charlesbank for children's gardens. Twice a week during the

planting, cultivating, and harvesting seasons, two processions of fifty children each can be seen marching, two by two, through the streets of the West End to their gardens. Over their shoulders, like a soldier's bayonet, are carried those worthier weapons, the tools by which human society has built its fabric—the hoe and the rake.

For the parents of these children some municipalities provided "garden city plots." In Minneapolis in 1912, 150 acres were cultivated on vacant lots. When growers complained about the overabundance of local produce, supporters cited improved sociability, health benefits, savings in food costs, and relief from the tensions of urban life in defense of community gardens.

With the world in crisis during World War I and as the farmers of the United States and Europe went to war, the governments of both England and the United States sponsored War or Liberty Gardens to offset food shortages. An American organizer of the World War I War Gardens effort later explained:

> The war garden was a wartime necessity. This was true because war conditions made it essential that food should be raised where it had not been grown in peacetime, with labor not engaged in agricultural work and not taken from any other industry, and in places where it made no demand upon the railroads already overwhelmed with transportation burdens. The knowledge that the world faced a deficit in food, that there existed an emergency which could be met only by the raising of more food, was apparent to every well-informed thinking man and woman during the early months of 1917. The idea of the "city farmer" came into being. . . . Near every city were vacant lots as useless as the human loafer. . . . As the campaign progressed, it was found that the best results could be obtained by organizing communities.[4]

In 1918, the U.S. War Gardens produced over 264,000 tons of fresh vegetables in 5 million gardens.[5] Children, too, were drafted into the community gardening movement with the creation of the U.S. School Garden Army. After the war, gardeners were urged to continue production for several years until conventional agricultural systems were brought back to normal.

Again, during a crisis, this time the Great Depression of the 1930s, Americans turned to community gardens. "Relief gardening" reappeared as a means of supplementing tight budgets and retaining the traditional work ethic in an era of high unemployment and economic upheaval. Similar to the community gardens that had been formed after the Panic of 1893, relief gardens produced staple crops of potatoes and beans on large 50-by-100-foot plots.[6]

But while England's allotment and small-holdings gardens were permanent, the community gardens of the United States were subject to whim, fashion, and favor. By the time World War II broke out, most of the older community gardens had reverted to weeds, development, or parkland. Once again, the Department of Agriculture, the public schools through PTA committees, parks departments, and civic groups began to plow up the earth on all available arable land near centers of population. Produce was consumed by local communities or sold, and savings were sent on to aid the armed forces overseas.

In many parts of the country, demand for garden plots far exceeded supply, and gardeners were chosen by lottery. Newspapers printed columns on gardening for the novice, and school and factory gardens flourished once again. At peak production in 1944, 20 million victory gardeners grew 40 percent of the fresh vegetables consumed in the United States.[7]

Once again, as the country's wartime atmosphere was replaced by prosperity, community gardens reverted to their former status as vacant lots, lawns, and parkland. The "American dream" of a single-family house in the suburbs, encouraged by cheap mortgages and extensive highway construction, dealt the final blow to most of the surviving community gardens. Even in Europe, where wartime gardening had been reorganized into a "leisure gardening" movement, the demand for plots dropped and weeds spread out across many of the permanent small holdings.

The environmental movement of the late 1960s and early 1970s, combined with sharp increases in food prices and concern about chemical additives in many processed foods, renewed Americans' interest in home-grown produce. During the same period, after enactment of a 1965 law opening immigration to Third World and other non-European immigrants for the first time since 1924, the United States experienced an influx of new residents from agrarian cultures. By the mid-1970s, these forces had combined to create a community gardening movement across the nation, largely by people whose rural roots allowed them to understand precisely what it takes to make a garden grow.

Almost all of our country's population has come from once-rural peoples who brought with them their own agricultural traditions. Such traditions die slowly, be they from Northern Europe, Africa, Asia, the Mediterranean, China, the Caribbean, or Latin America. The community gardens blossoming today reflect these rich and varied cultural patterns.

In the mid-1600s, the founder of Rhode Island saw the Narragansett Indians turning the sod in the spring and wrote: "With friendly joyning they breake up their fields. They have a very loving sociable speedy way to dispatch it. All the neighbors, men and women, forty, fifty, a hundred etc., joyne and come in to help freely."[8] Once prepared, the ground was divided into family plots. In our community gardens three centuries later, this would be a familiar sight—except that now, instead of from a single tribe, the neighbors come from all corners of the earth.

Part II

———◆◆◆———

FROM IDEA
TO
REALITY

3

Basic Steps of Organizing
a Community Garden Group

— ◆·◆ —

Judith Wagner

— ◆·◆ —

When Ed Cooper retired from his job at age seventy, he was just beginning a new career as a community garden organizer. Many elderly people lived in the area where Ed worked, and he found that four hundred senior citizens were living in his own neighborhood, the Highland Park area of Roxbury. Most of them, like Ed, were originally from the South and had grown up on farms. Having decided to keep busy in his retirement and hoping to improve the lives of people who, as he observed, "would spend practically all day just looking out the window," he formed a cooperative society for these elderly neighbors, the Highland Park 400 Club, from which a community garden emerged. Today the garden thrives, and Ed, who is in his seventy-seventh year, serves as president of Boston Urban Gardeners. He notes, "Gardening is one of the best avocations for old people . . . keeps them busy seven months of the year . . . makes them feel productive and they can share stories with people who have much in common." For the past five years there

19

have been eighteen garden plots in Ed's garden. The retail value of food produced has been estimated at $4000 to $5000. The garden group takes the position that gardening means "dollars and sense."

Ed Cooper's garden was organized to serve the needs of the elderly in his community, but many other needs can bring neighbors together to form a community garden. Isaac Graves, aide to U.S. Senator Paul Tsongas and another Boston community garden advocate, responds to a different need: "Boston, as an area, is a very small city, and abandonment, arson, and resulting vacant lots could have a devastating effect on a neighborhood, and it only takes one or two before it starts looking like gapping teeth in a smile."

For whatever reasons a group decides to form a community garden, the organization process is basically the same. The first step is to find out what needs exist. In the case of Boston, unfinished urban renewal projects (including a major interstate highway that was stopped by community protests in the early 1970s) left acres of empty, scarred land that contributed to a negative community image and led to the withdrawal of financial and political support in several of Boston's neighborhoods. There was a need to put this land to some productive use and to stop the dumping and abuse that disfigured the community. From experience we found that a garden was excellent evidence of community investment. When we also discovered that families could produce significant amounts of food at low cost, it was clear that the gardens would meet critical food and nutrition needs of the community. Over the years we have learned how gardens work to involve people of all ages across social, cultural, economic, and racial lines; we have observed cohesion and interaction stimulated among neighbors who had never spoken before they began to garden together. Now, when we approach a group starting a garden, we describe how a garden can meet any or all of these needs: needs for positive land use, for better food and nutrition, for community involvement, and for cooperative action. If people feel a need for any of these things, we know there is a good base from which to start the organizing process. A garden will not be successful if the people working on it do not feel it is doing something important for them.

The basic elements required to start a garden are: interested people; a suitable site; and basic resources—including fencing, topsoil, and water—at minimum expense. This section will cover the first of these three elements.

FINDING INTERESTED PEOPLE

Start with yourself. Whom do you know—neighbors, family, friends, acquaintances—from church, school, or local community centers who might be interested

News Notes

Time to Sign-up for Garden Plots

It's time to sign up for garden plots at the Southwest corridor Community Farm. The first meeting of the season will be held on Sunday, March 22, 2-3:00 p.m. at the Farm office, 46 Chestnut ave., Jamaica Plain. If you can't make this meeting or want more information you can call the Farm on Tuesdays and Thursdays at 522-1259. Residents of the Hyde Square area are particularly encouraged to think about community gardening at the Farm this year.

Interagency Breakfast at Little Wanderers

The next Jamaica Plain Interagency breakfast will be sponsored by the New England Home for Little Wanderers, 161 South Huntington avenue, Jamaica Plain, at 8:30 a.m. on Tuesday, April 21. Anyone who represents a social agency that serves JP or is interested in attending the meeting is welcome. Further questions may be directed to Michele Ingemi or June Leonard, 232-8600.

Advertising garden plots in a community newspaper.

in a garden? Talk the idea over with people you know so that you find support from the start.

Reach out to others. Ask your friends to talk to friends. Get a couple of friends to help you pass out a flyer to people in the area where you would like to garden. Don't be too shy to knock on doors and introduce yourself to your neighbors. Remember to use bulletin boards in community centers, laundromats, corner stores, churches, and other familiar places.

Place a news note (a short paragraph will do) in your community newspaper stating your intent to start a garden, a first meeting time, date, and place, and a contact phone number if possible.

Place a free public service announcement with a local radio station giving the same information included in your news note.

Be prepared to keep trying. Sometimes a first meeting will be so large that you will have trouble handling all the interest. More often the first meeting will be small. Don't be discouraged. Use each meeting to build interest for the next. The people who come to the first meeting will probably be the optimists who are willing to work hard. They may agree to form a planning committee for the garden. Even if only a few people come, encourage everyone to bring at least one other person to the next meeting. The initial meeting should be used to verify the needs of the community, to take an inventory of any resources people know of, and to set up a strategy for creating the garden.

Remember that it takes only a few people to provide the basic resources for the garden. Once those are established, it is our experience that this proof of progress will impress the more skeptical people who will soon be drawn in.

THE FIRST
ORGANIZING MEETINGS

The first several organizing meetings should focus on acquiring the basic elements for the garden. Organizing is usually needed in two categories: people and resources. It is important to expand the list of contacts and people who might want to join the project. It is also important to list the resources required to start a garden—land, topsoil, fences, seeds, tools, water—and how they can be obtained. To keep the meetings productive, try the following methods:

1.   Have in mind the reasons you feel a garden will be good for your neighborhood and some examples of successful gardens that can provide a model for your own. If there are no examples close at hand, knowledge and photos or descriptions of gardens in other areas or cities can be equally inspiring.

2.   Prepare information about sections of empty land within the community that may make good garden sites, and check the possibilities of your using them.

3.   Take an inventory of the resources represented by or known to the people at the meeting. Who knows how to write up a flyer? Who is handy with tools? Who has time to search for a site? Who can do plumbing work?

4.   Spell out the tasks needed to locate a site and to get basic resources, then assign each task to volunteers. Encourage everyone to take part in this.

5.   At each meeting be sure to get reports on all the assignments from the last meeting.

6.   Be prepared to provide encouragement to people who run into difficulty completing their tasks. Sometimes teamwork can make it easier to solve problems. It is useful to team up someone with experience with others who may be new to such work.

Providing a "critical mass" of resources may be essential in getting beyond the initial organizing steps, especially in communities that have been subjected to many failures or where people have become convinced that they are powerless to change anything. A critical mass can consist of information, specific resources, or even positive stories that convince people they are not alone and *can* get enough materials and energy together to do what they want.

You, as a garden organizer, must do your homework in order to offer specific, credible ideas for solving the problems facing the group. The organizers must always be open to ideas, suggestions, and information provided by other people during the meetings, but without the basic information at hand many meetings can be quickly stalled.

THE GARDEN ORGANIZATION

Once the garden site is secured you can begin to expand your group. We found that when we delivered topsoil to a site it was easier to attract people because they could see something happening. People asked questions as they passed by; and we painted a sign designating the site a community garden and providing a phone number for people who wanted more information. More meetings can be called, but the most effective organizing method is action. Once skeptics see others working and getting plants to grow, they become more willing to participate.

At this stage it is important to have one or more people clearly designated as coordinators. To begin to spread the responsibilities around, it may be useful to consider people who were not the original organizers. Two people may want to share this job, although more than two may cause confusion. A coordinator should live near the site, or at least be easily available so that people do not become frustrated when they try to learn more about joining the garden. A coordinator should be the type of person who is willing to work on solving problems as they arise and who will not be easily put off or take "no" for an answer. The original organizers of the group can, ideally, remain in the role of supporters and provide various sorts of back-up assistance to the coordinator as needed.

As the number of people involved in the garden increases and as the core group expands, there will be a need to develop the structure of the garden and the rules for setting it up. Each garden group should discuss and choose its own design. A number of questions need to be answered:

1. What are the conditions for membership (residence, dues, agreement with rules)?
2. How will plots be assigned (by family size, residency, need, or group, i.e., the young, the elderly)?
3. How large should plots be (or should there be several sizes based on family size or other factors)?
4. How should plots be laid out?
5. If the group charges dues, how will the money be used, and what services, if any, will be provided to gardeners in return?
6. Will the group do certain things cooperatively (such as turning the soil in the spring, planting cover crops, or composting)?
7. When someone leaves a plot, how will the next tenant be chosen?
8. How will the group deal with possible vandalism?
9. Will there be a children's plot?
10. Will the gardeners meet regularly? If so, how often and for what purposes?

11.   Will gardeners share tools, hoses, and other such items?

12.   How will minimum maintenance (especially weeding) be handled both inside plots and in common areas (such as along fences)?

13.   Will gardeners be expected to uphold a set of written rules? If so, how will they be enforced?

14.   Should your group incorporate and consider eventually owning your garden site?

Remember, there are no right or wrong answers to these and other questions your group will think of. But discussion and planning will help you create a garden organization to suit your needs.

In many cases these questions will be answered only with time and experience. But discussion, particularly of the criteria for obtaining a plot and standards for plot maintenance, will be important right from the start. Some issues that may take some research include whether gardeners should be allowed to use chemical pesticides and fertilizers and how plots should be laid out to prevent one person's plants (or a tree or building) from shading another's.

Once a group of gardeners begins to discuss these questions and the first crops are planted, a garden can be considered "organized." Then energy must be turned to the maintenance and support stage of the garden's development.

A garden plot application can spell out rules of the community garden and provide a mailing and phone list of gardeners each year.

A P P L I C A T I O N

Application for a single plot ( ), double plot ( ) in the
_____ . Single plot size is approximately 9′ × 15′.

Number of individuals being fed by the plot: _____

Number of years of gardening experience: _____

I understand that I must comply with the following guidelines in order to keep my plot for the following garden season:

1. Maintain a neat garden, keeping garden area and surrounding paths free from overgrown weeds.

2. Remove all unsightly piles of rocks or stones or other forms of litter into trash barrels. Waste vegetable matter shall be placed in compost bins.

3. Harvest all produce when it has reached maturity (or arrange to have it harvested to deter vandalism).

4. Remove all plants when they have ceased to bear.

5. Maintain garden consistently throughout the growing season and arrange that care shall be taken should extended absences arise, such as vacations.

6. Comply fully with all federal and state laws and regulations regarding the use of pesticides and other chemicals.

7. Respect all other gardener areas, and not trespass upon other garden plots.

8. Participate in the vandal watch, report incidents of vandalism to the garden coordinator, and get to know fellow gardeners.

9. Make concerted efforts to conserve water by turning off faucet after watering garden.

10. Cars will not be allowed past black-top area.

Signature _____ Telephone (home) _____
Address* _____   (work) _____
_____

*Please report any change of address   Original to garden file
 to garden coordinator   Copy to gardener

4

Resources:
What You Need and
Where to Find It

———— ◆•◆ ————

Greg Watson

———— ◆•◆ ————

Successful community gardens rely on human and material resources almost to the same degree. Gardens don't develop without gardeners, and gardeners aren't likely to be successful without topsoil, water, seeds, and basic tools. The first task that a group of prospective community gardeners should undertake is to find out what the group needs.

Table 1 (page 163) lists most of the basic material needs and costs associated with establishing and maintaining a community garden. This table includes estimated expenses, many of which will not be applicable to your particular garden. For example, some of the resources such as topsoil, water, or even fencing may already be available at your garden site. The prices listed range from the minimum costs of a small garden (5,000 square feet) using the least expensive materials and donated labor, to the maximum costs for a large garden site (40,000 square feet)

26

using high-quality, new materials and contracted labor. It does not include any staff or administrative salaries.

All of the items marked with an asterisk can be found for free. Methods for obtaining them will be discussed in detail later in this chapter. Notice that there are only a few articles that are extremely difficult, if not impossible, to obtain free of charge. A little resourcefulness, however, will go a long way toward reducing your group's expenses.

Make sure that you do not overlook or underestimate the information and skills of your gardening group. You may already have experienced gardeners who will be helpful in assessing the garden's needs.

HUMAN RESOURCES

Every community garden should have a coordinator or facilitator. The coordinator must be willing to put in extra time—usually on a volunteer basis—to coordinate the activities of the gardeners and the operation of the garden. In their unique role, garden coordinators have been variously described as saints, martyrs, angels and fools.

The tasks the garden coordinator can expect to manage include

- Locating a garden site
- Finding a sponsor for the community garden (if desired)
- Making certain that critical resources such as water, fencing, and plowing are available
- Seeing to it that the financial matters of the garden (fees, fundraising, grants, etc.) are taken care of efficiently
- Recruiting gardeners and publicizing the garden project
- Organizing garden meetings
- Organizing plot layout and assignment
- Encouraging gardeners to become involved in all aspects of the community garden (especially in deciding the structure and decision-making processes of the group)[1]

The garden coordinator is critical as a catalyst and must organize the gardeners into a cooperative and harmonious group. The importance of the last item on this list of responsibilities cannot be overstated. Community gardens *are* cooperatives, although we often don't think of them as such. The dictionary defines a cooperative as "a jointly owned means of production or distribution of goods or services operated by the consumers for their mutual benefit."[2]

Specifically, community gardens are producer co-ops. But the important word
here is *co-op*. Although it is important to have a garden coordinator who can provide a certain measure of leadership for the project, it's even more important to
involve everyone in the group with the garden's organization from the start. This
is essential for preserving the unity of the group and the sanity of the coordinator.

NETWORKING

An effective network is the single most valuable "tool" that a community gardening
group can have. What's more, it's accessible to any and all groups free of charge.

Networks are "individuals and groups with similar concerns and complementary
resources who are 'linked together' in collaborative efforts as well as in sharing
information and support."[3]

There is nothing complicated about networks. Many are casual and informal.
A network is usually created in response to a particular need, often without any
formal recognition of its use. In fact, many people who participate in networks are
not even familiar with the term. For example, if you've ever helped a friend find
an apartment or job, you've participated in a temporary informal network, that is,
a network with a specific goal that, once attained, usually signals the termination
of the network.

Boston Urban Gardeners is part of a citywide network of individuals, groups,
and agencies in the Boston metropolitan area who understand the importance of
urban agriculture and are committed to helping in the "greening of Boston." The
Boston Urban Gardening network includes individual gardeners and garden
groups from virtually every neighborhood in the city. In addition to this grass-roots
base, the network has representatives from the federal, state, and private business
sectors as well.

Directly or indirectly, this vast network, facilitated for the most part by BUG
staff and volunteers, has been instrumental in helping gardeners locate and obtain
everything from garden sites and topsoil to garden tools and garden coordinators.
Of equal—if not more—importance is the moral support and shared sense of purpose communicated through such networks.

Generally speaking, the size of a particular network is determined by the
breadth of the goals of the participating members. If a group's interests are strictly
garden-related, its network will probably be composed mainly of other gardeners.
Some networks' main objectives, though, might be sharing information and sup-

plying moral support, and others might wish to embrace larger, more specific goals. Members might feel strongly about emphasizing the ecological benefits of organic gardening, or about educating the general public on the need for open agricultural spaces in the city. Support for these kinds of causes will certainly draw the attention of other individuals and groups who share these concerns. Networks come into being by being used. They save gardeners time, aggravation, energy, and money.

SCROUNGING

scrounge (skrounj), v., *scrounged, scroung-ing,* n., 3. *scrounge around,* to search or forage for something, especially in a haphazard or disorganized fashion; hunt for; We scrounged around for something to eat.[4]

If there's one resource a community garden group can count on to be scarce, it's cash. Therefore members are often challenged to come up with creative strategies for obtaining material resources at little or no cost. As a result, the word *scrounging* has taken on new meaning. When used in the context of community gardens, scrounging is an effective way to recycle and conserve the earth's resources.

Scrounging means different things to different people, but each of the following activities can be considered a form of scrounging:

- Repairing and/or rebuilding used equipment and tools
- Constructing tools from recycled materials
- Soliciting donations of materials, money, services, or time from local businesses, organizations, or individuals
- Hunting around and buying high-quality secondhand tools and materials at yard sales or flea markets
- Inventing new uses for throw-away materials such as barrels, plastic jugs, and wooden pallets (e.g., to create water systems or fences)

Two qualities are essential to successful scroungers: resourcefulness and persistence. A good scrounger always remembers that it is impossible to exhaust *all* resources.

The single greatest limiting factor for successful scrounging is often transportation. Once you discover good places for locating recyclable materials, you may want to team up within your garden group to be sure that people without transportation will have access to the available resources.

Recycled cobblestones provide a protective border around a newly planted tree.

Table 2 (page 165) offers a variety of suggestions for scrounging possibilities. Fortunately—and yet unfortunately—this is a wasteful society, and materials for recycling abound. It does take some effort to get into a scrounging frame of mind, to see the potential, if unorthodox, uses for materials or sources for something you need. Once you get into the spirit, though, you'll see opportunities at every turn.

SOURCES OF FUNDING

After assessing the needs of your garden group and how many of these you expect to obtain for free, your group will be able to estimate how much money it will have to raise to purchase the remaining materials or services. Depending on a number of factors, this figure could be either relatively insignificant or terribly prohibitive. If the former is the case, expenses might be met by collecting a small fee from each gardener. If, on the other hand, your group has minimal access to community resources and is faced with the prospect of having to purchase a substantial amount of materials and services, a combination of strategies might be undertaken as shown in Table 3 (page 167).

Your group might consider putting on a fundraiser of some type. Fundraisers range from potluck dinners to car washes, raffles, or dances. If your group does decide on a fundraiser of some sort, weigh as many factors as possible before choosing the activity. Before deciding upon a dance, check to see if you can obtain the use of a hall and the services of a band for free. If these become expenses, chances are your dance won't raise much, if any, money for your cause. On the other hand, Boston Urban Gardeners found eighteen gardeners willing to sew a patch for a quilt that served as a unique grand prize in a raffle. Expenses for this kind of fundraiser are minimal.

Finally, your group might attempt to tap the resources of government and private foundations. Cities and towns often receive federal and state funds for distribution as part of "revitalization" projects or block grants. In some instances, garden groups might qualify for such funding. A list of foundations in your area is usually available through local banks, churches, or the public library. Table 4 will help you organize your request for funds. Although these requirements for writing a funding proposal may seem discouragingly formal and difficult, remember that the purpose of such a proposal is merely to tell your story clearly and simply. If you do just that, you should be in a strong position to receive a positive response.

An important last point to consider if you find that you must buy some tools and materials is cooperative purchasing. This may be appropriate for expensive tools, such as wheelbarrows, as well as for materials needed in large quantities, such as straw for mulch. Although cooperative purchasing necessitates some additional time devoted to scheduling tool use, in the long run it can save the group a considerable amount of money.

Your community garden group should explore as many of the options described here—along with others you come up with—as time and energy permit. Encourage each gardener in the group to share the responsibility of thinking up ideas for new resources. Garden members, after all, are resources in themselves.

5

The Public Role

———◆◆———

Susan Redlich

———◆◆———

What's the greatest problem? I think it's the attitude that somehow the city administration should take care of open space for us. And I think we're never going to have all the open space we'd like unless each of us who is at all interested is willing to pitch in and do something, perhaps individually, even more importantly and more usefully, with a community group that will say we'd like to take over this parcel and we will take care of it. . . . There's never going to be enough city money to hire people to pick up after all of us, and we have to do that ourselves.

This sentiment expressed by Isaac Graves, aide to U.S. Senator Paul Tsongas, who has worked in both community and government agencies at all levels, should guide you in your search for public support. Although it is appropriate—and your right as tax-paying citizens—to solicit public support for community garden development, your group will inevitably be stronger if it can draw what it needs from its own ranks.

Before searching outside your garden group for assistance, it is important to recognize the skills available within the group. Jot down your group's objectives, and take stock of what you already know. Get together with your group, even if it's just you and your next-door neighbor at this point, and see if you can fill in your own version of this sample chart.

32

What do we want?
1. Ownership of the garden lot
2. More gardeners
3. Fencing
4. Better production

What resources can we tap within our group?
1. A lawyer
2. A super gardener
3. A truck driver
4. A real estate salesperson

What do we need?
1. Money to purchase the land
2. A lease
3. A pick-up truck
4. Books, sources of gardening information

Where and how can we get assistance?
1. City Real Property Department
2. State Department of Agriculture
3. State Community Development Office
4. Municipal Parks and Recreation Department

When you reach the last question you'll be ready to start knocking on doors, better prepared to get what you want, and better able to describe and find the assistance you actually need.

STALKING THE GOVERNMENT

Public agencies may be the first or last place to turn, depending on the type of problem you face. Federal, state, and local agencies that may be helpful include those that deal with agriculture, business development, community services, and education. In general, these agencies can offer the most help in the area of information services and referral. In the case of community gardening, they may be helpful as well in supplying equipment and material for the garden site. When approaching an agency, you can save a lot of your time and theirs by asking three questions:

1. Do you have the information I'm looking for (names of other garden groups in the city, pamphlets on how to clear a vacant lot, etc.)?

2.   Can you send me the pertinent materials (application for land lease, agricultural agency addresses, etc.)?

3.   What other agencies might be of assistance to our efforts?

You should be able to increase your rate of success by starting early—at least several months—before you actually need the assistance you are requesting. When you suspect that more than a phone call will be needed to spring the help loose, write a letter outlining the nature of your request and follow it up with a personal visit.

The challenge in soliciting assistance will be to distinguish your requests or demands sufficiently to attract the attention of the people who can help meet them. Several tactics should be used.

First of all, you should make an effort to collaborate with other groups in seeking help. Two garden groups in the same neighborhood will command more attention together; and economies of scale might be achieved, for example, sending out a front-end loader to your part of town to clear two lots in one day is more efficient than filling two separate requests.

Document the benefits to be realized from your project, using numbers and facts as much as possible. Mention, for example, the amount of produce to be grown, and cite statistics on the number of people it will serve. Show the agency how their help will go a long way. For example, a little material support in the form of plastic water lines is a one-time request that the gardeners can maintain for years.

Present yourselves as much as possible as a self-help effort. Most public grants are contingent upon the recipient's willingness to provide a match of money, labor, or demonstration of self-sufficiency once assistance has been given to the project. To do this, highlight your history (if you have one) of progress. Mention the people and organizations that endorse your ideas, such as the local state representative or the senior citizen council. Even if you haven't actively worked with these groups, you can gain their support through a simple presentation or through correspondence. Also, generate publicity surrounding your efforts—human interest stories in local papers, photographs of people pitching in, and any other kinds of public recognition of the project.

If you approach an agency that is at a loss to provide you with direct assistance, ask how they can help you with your request to another agency or organization. This might mean their writing a letter of support, telephoning a key individual, or promising supplemental help (such as an easement in the deed of sale if the group can raise the money to purchase land at open-space price).

Obtain a memorandum of understanding if you can. Although not a binding contract, it is better than a verbal commitment. The memorandum of understanding is a pledge of mutual responsibility, not involving money. For example, a memorandum of understanding could secure access to land, renewable yearly as long as certain conditions are observed.

Bargain for services. In these days of budget cutbacks, nonmonetary exchanges or barter might get you what you need. For example, in exchange for plowing services from the adjacent prison farm, a youth garden group plans to grow a few extra rows of cucumbers and tomatoes for the institution's kitchen. In exchange for donating a load of manure to a tax-exempt school garden, a farmer received a receipt for the value of his donation, which he could use as a tax write-off.

Another strategy you can pull out of the hat if it's timely is to ride on the coattails of another project. If public funds are being spent to tear up the landscape in some fashion, say, for a housing development, a garden could be created while the earth-moving equipment is on site. Armed with information on the need for garden land in the area, your group will be in a better position to negotiate claims to space and appropriate designs. Or you may be able to join forces with another publicly funded program. A patients' garden at a hospital can become a joint venture with the community; a school garden stands a much greater chance of survival over the summer when it's part of a community garden; an energy-saving greenhouse built into a public building could be sure of maximum utilization and management if a community group is involved with it from the start.

A rule of thumb in dealing with agencies is to plan a shopping trip with the idea of packaging the help you obtain. When planning this expedition, think about how you can define the goals of your project within the framework of what that particular agency or government office does or has to offer. The following subject areas may give you some ideas of ways to think of community gardening.

Community Development

As a project that calls forth cooperation among, and self-help efforts by, neighbors, community gardening certainly qualifies as an effective vehicle for community development.

Resources to tap: local government grants (via city hall), local chamber of commerce, local merchants' associations, other city and state agencies dealing with community development, national agencies concerned with local development, private foundations, local banks and corporations (c/o Community Relations Departments).

Education

People of all ages benefit from the garden experience. It is a learning tool for everything from horticulture to equipment repair and basic job skills.

Resources to tap: state department of education, boy and girl scout organizations and clubs, county 4-H programs, local school departments, private foundations.

Recreation and Therapy

Many community recreation programs include gardening as an activity to help build individual pride and self-confidence.

Resources to tap: local recreation departments, mental health agencies, senior citizen centers, horticultural therapy associations.

Agriculture

The most obvious skills that gardening calls upon are agricultural.

Resources to tap: county cooperative extension service, state department of agriculture, conservation district (through the nearest U.S. Department of Agriculture Soil Conservation Service), vocational agricultural schools, nearest arboretum organization, horticultural association or garden clubs, manufacturers of agricultural products.

Nutrition

The harvest of fresh produce is a low-cost way to bring nutritional benefits to city dwellers.

Resources to tap: city and state nutrition services, county cooperative extension home economics staff, nearby university nutrition departments, supermarket corporations, public health departments.

Beautification and Conservation

The before and after comparisons of a vacant lot turned garden tell the story.

Resources to tap: horticultural societies and garden clubs, local and state conservation commissions, local newspapers, the city parks department, local neighborhood associations, men's and women's clubs.

Energy Conservation

Cite the amount of oil used to get a bunch of carrots packaged, shipped, and delivered across country in contrast to the amount used in community gardening.

Resources to tap: state energy office, U.S. Department of Energy, state department of transportation, private foundations.

Economic Development

Time and labor invested in a garden yield high returns in the form of food dollars saved ($950 per thirty-by-forty-foot garden plot if intensively cultivated), job training in horticulture and landscaping, and possible commercial ventures (seedling sales, farm stands, produce sales to local institutions).

Resources to tap: city and state economic development agencies, community economic development corporations, local food cooperatives, local chambers of commerce.

When contacting these various sources, consider the widest range of assistance they might provide—information, in-kind services, such as consultation, use of a telephone or office space, typewriter, mimeo machine, truck, plant stock—as well as money grants.

STEPPING INTO THE POLITICAL ARENA

Each time you organize to meet community needs you engage in political action. Some community problems are more easily solved if people or groups act jointly rather than individually. Combined with demonstrated self-help, the joint approach can often attract government support as it shows strength in numbers. Some situations may require political involvement at several levels in order to influence a decision in your favor. To affect decisions that involve the use of public resources such as land, you may have to get support from municipal officials as well as state agency directors, state legislators, and even congressional delegates. These elected officials and public employees are much more likely to respond to an organized coalition of groups than to one or two alone.

One example of this strength in numbers occurred recently. Two hundred acres of sparsely developed, state-owned land located within three miles of downtown Boston was no longer needed by the state mental health hospital. A ten-acre portion of the site currently hosts approximately four hundred community garden plots that produce over a quarter of a million dollars' worth of vegetables for home consumption. A citywide coalition of garden groups has made open space an issue to be addressed by anyone designing plans for the land's reuse. These groups are promoting the concept of a "productive landscape." Components of such a landscape, which might include an agricultural preserve, a greenhouse, a compost facility, and a landscape nursery, have been discussed. Even the redevelopment of the site for housing and industry could be designed to include these projects.

Members of the garden coalition have used every opportunity to promote this concept of a productive landscape. At an awards ceremony for one of the garden groups, the group used the occasion to ask the governor for his support. Concurrently, a bill was filed in the legislature to dedicate a percentage of the acreage as an agricultural preserve. Community gardeners testified at the hearing. The state's Division of Agricultural Land Use obtained assistance from graduate landscape architect students to develop a comprehensive landscape plan emphasizing the agricultural potential of open space in urban areas. The preparation of the report involved discussions with representative community groups in the adjacent neighborhoods, city and state officials, and local planners. At a one-day participatory design workshop, gardeners engaged in brainstorming to come up with ideas for the site. The report and accompanying slide presentation would be used to increase community support for the plan and to persuade the political and economic interests to incorporate these criteria in any future schemes for redevelopment of open space.

Never underestimate your ability to influence people. A woman testifying at a land-use hearing stated, "I've never done this before and I'm nervous as a kitten. I'm begging you, don't take away our garden." She burst into tears but quickly regained her composure and went on to give an eloquent appeal for her garden, which was located on state-owned land. Her message was so moving that the legislators and other witnesses applauded at the end.

The citizen action in these cases is already stimulating action from other groups located near surplus public properties across the state. This process has generated ideas for open-space land use that are receiving the attention of public officials, perhaps for the first time. As such it sets a precedent for future public land redevelopment.

CHARTING A COURSE FOR COMMUNITY GARDENING

Petitioning government agencies for an ever slimmer piece of the pie diverts people from what could be a more powerful course of action. Political involvement can help build public policy favoring community gardening without necessarily competing with other interests. As expressed by one Boston community gardener, Esther Diggs,

> If you're worrying about a vacant house, you're worrying also about a vacant lot that that house could become. There's no way you can separate those issues. . . . We're talking about the whole quality of life, and open-space issues are part of that and they can't be taken away. People need housing, people need recreation facilities and so forth, and people need greenspace [and] a garden.

By looking at the larger context, community gardeners can join a wider community of interests connecting people across and between neighborhoods and entire regions.

Has your city or state government expressed a policy in favor of community gardening? If so, you can use that policy to give weight to your proposals. When gardens are pitted against some contrary proposal, such as a proposal for a parking lot, you can remind officials of the government's stated commitment. Where will you find evidence of such a policy? Check out your city's municipal open-space plan. Every city has been required to have one in order to apply for federal recreation and conservation grants. Talk with the local conservation commission to learn how it stands on issues involving the use of open space. Ask the state department of agriculture what policy it follows, and be sure to investigate the legislature's committees on natural resources, food, and agriculture. Statutes may already be on the books to promote community agriculture.

The U.S. General Services Administration now has a management policy regarding its properties called the Living Buildings and Community Gardening program, which allows community gardening on available sites when feasible. If there is federal property in your area that could serve as garden land, contact your regional GSA office, attention of the public buildings service, and state your proposal.

If there are no explicit public policies, progress in community garden development stands the risk of being outflanked by competing interests for land. How does a policy become established in the first place? In order for any policy to gain acceptance and be implemented, people who care must participate in shaping the major tenets of that policy.

The power to initiate this process and to rally public support behind it for policy adoption can come from a coalition of interests. A coalition of separate groups with common interests can serve as a forum for bridging minor differences and arriving at a mutually agreeable position. From there the various groups can mobilize their members to organize the support of others.

Several cities around the country have addressed food issues by developing city food plans that take an integrated approach to solving urban food problems. In Seattle, the Department of Community Development saw the possibilities of projects like the Pike Place Market and Bulk Commodities Exchange to grow into programs such as community greenhouses, educational workshops, and a food-processing center. One report states:

> What was needed was an overall plan that included feedback, input, and recommendations from citizens. The department's staff took the initiative to draft working papers; to set up an Urban Agricultural Technical Advisory Committee; to receive recommendations, and finally, to submit policy recommendations. . . . These policy recommendations were reviewed by citizens and then approved by the mayor and City Council. This process in turn has laid the groundwork for Block Grant funding for the recently incorporated Neighborhood Technology Coalition . . . which will be able to make small grants to projects related to food, housing, energy, waste, and water issues.[1]

Consider various strategies for building a strong policy. Forming alliances and coalitions to discuss common interests can be a starting point. Your group can call hearings, form task forces, and circulate petitions. Present your legislators with a resolution that you want them to pass. Stage a public event where the governor, or other prominent political figure, can make a declaration of support for community gardening. It is even possible to unite a coalition that spans urban and rural interests. One successful group in Massachusetts has accomplished this. The Massachusetts Food and Agriculture Coalition (Mass FAC) works on a statewide level to protect food production lands and community food efforts. Mass FAC informs its members of important pending legislation through action notices, urging each member to express an opinion. The group has also adopted policy plans through regional and statewide conferences. These policies form the basis for the group's support and sponsorship of various bills. Several legislators have joined the group, in fact, and assist in finding effective approaches to various policy questions.

Support your policy goals by citing facts. Massachusetts legislators receive a Food Survival Kit every spring, with information concerning food issues in the commonwealth. Print up your recommendations in a flyer or in newspaper form, and distribute it for discussion. Through involvement and contact with the wider public, you can begin to influence the outcome of political decisions.

You *can* make a difference. Most state legislators and municipal officials hear so little from their constituents that a few phone calls and letters go a long way to impress them. Even if you have never tried to contact your public officials, you will soon learn how by asking experienced groups, the League of Women Voters, for example. A small amount of experience will quickly open the doors to many of the suggestions offered here. It may surprise you when a small group actually becomes involved in the democratic process and can work effectively toward convincing results.

Part III

---◆◆◆---

SITE SELECTION
AND
DEVELOPMENT

Photo by Read D. Brugger

6

Locating Land

◆•◆

Charlotte Kahn and Virginia Scharfenberg

◆•◆

Right now the way vacant lots affect people is that it's all weeds and grass growing on it and they're garbage collectors. . . . We have the interest and the people to have had four or five or six or more community gardens this year. As a matter of fact, people are taking over vacant lots—unfenced vacant lots—people are moving big chunks of concrete and the foundations off of them by themselves and planting without a fence. And they do without a fence because they organize the kids, and the kids help out in the gardens because they see the gardens as partly their own as well. And no fence, no vandalism. There's a garden right on Norfolk Street, which is a very busy street, that is surrounded by chicken wire four feet high with wooden poles, and it's a beautiful garden: a little, teeny vacant lot, the ownership is still in tax title, as are most of the lots, most of our 100 vacant lots on twenty streets in this area. And people know they're taking a chance when they start to use them and develop them on their own—they just can't wait, but they do it anyway.

—Tom Libby, We Can Neighborhood
Association, Dorchester

Finding a site for a community garden may not seem difficult if you live in a neighborhood similar to Tom Libby's, with 100 vacant lots in a twenty-street area. But to find the *best* site, even with such an abundance of land, requires a systematic approach that may be quite time-consuming. However, the time and effort is well spent if the end product is a beautiful and productive community garden that is important to your neighborhood's health and stability.

Because of the time and energy required to locate and acquire available land, a newly organized garden group should learn to select and develop a site using its

45

own resources: don't leave the site selection and planning to someone else. You, as residents, undoubtedly understand your neighborhood's dynamics and needs best and can determine what site will be best for your community garden. If you are thinking in terms of a permanent site especially, it is a good idea to think about the garden as part of a long-term plan for land use in your neighborhood. The following steps will help you to determine what is good land use in your community.

EXPLORE YOUR OPTIONS

If you are starting without a specific site in mind, you need to define convenient boundaries within which you will be looking for land. You will need the following:

- A good detailed street map of the area, usually available from your city or town planning department
- A minimum of three different colored magic markers and a couple of pens
- A camera (optional)
- A desire to work with your neighbors; a vision of how your neighborhood might look with a productive garden; perseverance and a sense of humor.

Begin by making a block-by-block survey. Catalog the buildings, open space, and vacant lots in the area. Draw a rough map of the area, and use your colored markers to indicate buildings, vacant land, and areas where people congregate or travel on foot regularly. Take notes on the uses to which the vacant lots are put, and estimate the amount of sun each lot receives. If a lot serves as a shortcut between streets or is otherwise being put to good use, you may not want to disturb it. On the other hand, if it's currently being used for illegal trash dumping or something equally undesirable, it's a prime site for a community garden.

Next you will need to research the ownership, square footage, parcel numbers, and property tax information on each potential garden site. This material should be available from the assessing department of your city or town hall. Ask a clerk for help in gathering this information, and keep careful written records for each site you hope to use. The research may not be quick or easy, but it's necessary whether you request assistance from a public agency, decide to purchase the site from an owner, or want to negotiate a long-term lease agreement. Some sites would be eliminated at this point if owners are not interested in your proposition or for some reason the site turns out to be unavailable.

EVALUATING A POTENTIAL GARDEN SITE

The next step in selecting a site is to evaluate each vacant lot's potential as a community garden. Before you invest time, energy, and money in developing a garden, it is worth finding the site most appropriate to your needs in terms of ownership, long-term use, accessibility to the group that wants to use it, and potential for supporting the kinds of plants and activities you have in mind. Following is a list and description of criteria for evaluating a vacant lot.

Soil Condition

Many urban vacant lots that have had houses or other structures on them in the past tend to have nothing even resembling soil on the surface. Grasses or clover growing on the site are indications of soil. Weeds alone, however, no matter how tall and luxuriant, probably indicate rubble. The worst rubble won't grow even weeds. If your site features nothing but the hard-packed remains of a former building—plaster dust, bricks, stones, assorted debris—imported topsoil is required. Before you order topsoil, take a half-cup sample of the site's surface to a scientific laboratory and have it tested for lead. It's possible that the old building's interior or exterior walls were painted with toxic lead-based paint and that lead paint chips remain on the site. If the lab test indicates that the site has a high lead level, you should either look for another site or take extra precautions when preparing the site for gardening. Such precautions would include gardening in at least twelve inches of imported topsoil and taking care not to disturb the original surface when turning the new soil for planting. Topsoil and soil amendments (peat moss, composted manures) are expensive and sometimes difficult to obtain, and you can use them most efficiently in carefully prepared raised beds (see Chapter 13). Paths between the raised beds can be covered with inexpensive crushed stone or wood shavings. To find nearby suppliers for all of these materials, use the Yellow Pages and compare suppliers' prices. You can then estimate your costs (see Table 1) before taking on a hard-packed site.

If the site does contain a layer of soil, dig several holes to see how deep the soil is. Eight or nine inches of soil is enough for a vegetable garden. Have the soil tested to determine its nutrient content and possible contamination by lead or other undesirable pollutants. Examine the site carefully to see whether it has been properly graded (is more or less flat) and whether the ground is uniform (is one portion an old black-topped driveway or cement foundation?). Consider the work involved to even out the site's surface and make it arable.

Slope

The slope of the site is important for several reasons. If the slope is too steep, rain will erode the existing or imported soil, which will accumulate at the lower end of the site. A steep slope may need terracing or special treatment. If the site slopes toward the north, it may be shaded much of the day. A gentle slope with a southern exposure would make a fine garden.

Fencing

If a site is already fenced, you will save a lot of money in its development. Remember when erecting a fence that trucks should be able to get into the garden (or as close to it as possible) with supplies such as soil, manure, railroad ties, and other large loads. Look for curb-cuts from the road, gates at least ten feet wide, and a driveway or other entrance that can be used to facilitate deliveries by truck.

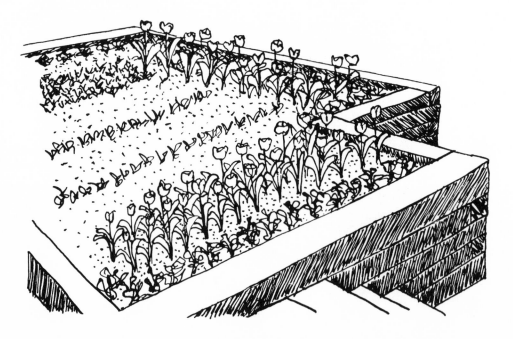

Railroad ties used to form a retaining wall and planting bed where the site slopes sharply to the street.

Sunlight

We all know that the sun rises in the east and sets in the west, but we forget that it shines from the south. A site with no adjacent obstacles, such as buildings or trees, on its south side will get sun all day long. A site bordered on the south side either by large trees, a tall fence, or buildings will be shady. Trees or buildings

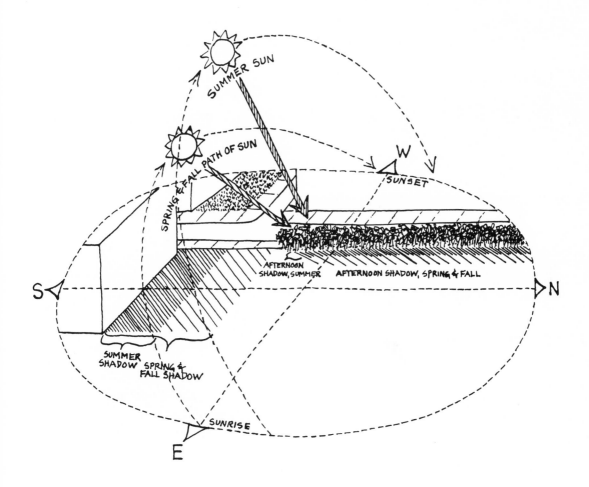

Following the sun's path across a potential garden site.

to the north will not shade the site, but tall obstacles to the west or east will create partial shade on each side during the course of the day, especially when the sun is low in the early spring and fall. You may need to check your future garden site at different times of the day to be sure of the available sunlight.

A vegetable garden needs six or more hours of direct sunlight a day to support healthy plants, but some varieties of food-producing plants, both vegetables and shrubs, can do well with less. Shady spots can also be beautified with plants that grow naturally in wooded areas.

Existing Vegetation and Natural Features

Natural features such as trees, rock outcroppings, hills, existing shrubs, and wildflowers can be special attractions to your community garden.

Access to Water

A source of water is essential for community gardening, especially if the garden's foundation is porous rubble through which water drains quickly. Possible sources of water should be searched out right from the start—a friendly neighbor with an outdoor faucet, a stream or river with clean water, a nearby fire hydrant. Chances are slim that former house lots will have a visible source of water, but most of these will have old water lines underground or within easy reach; water mains flow under most residential urban streets. To locate a water main, you can go to the local water and sewer department or to a similar municipal office and ask to see a water line map for your area. Be sure to check for any relevant regulations that may prohibit access to hydrants or water mains for safety or health reasons.

A site with relatively deep, friable soil will retain water and may be gardened using only regular rainfall and conservation techniques such as heavy mulching. You can test the drainage of the site by examining the ground after a heavy rain. If long-lasting puddles form, the site's surface is densely compacted and unsuitable for gardening without the addition of new topsoil or organic soil-amending materials (peat moss, compost, manures). Good drainage is essential for plants. Without it, their roots can drown or rot. On the other hand, a porous surface such as sand or the loosely filled foundations of an old building will drain too quickly to satisfy the water requirements of most plants. Daily watering might keep the plants happy, but it would be throwing time, money, and a precious resource literally down the drain. Organic soil amendments will both lighten heavy soils and improve the water retention of porous sites. Once you have developed a soil rich in organic matter, and if you mulch it heavily, watering should be necessary only occasionally.

But developing good garden soil takes time—sometimes years. If you have a choice, begin with a natural base of good soil.

Distance from Major Streets

Airborne lead pollution from automobile exhaust creates a problem for vegetable gardens less than 100 feet from a heavily traveled street. At distances greater than 100 feet, however, lead levels drop sharply. If you have a choice of available lots, the distance of each lot from main thoroughfares should figure as an important consideration in your selection.

Visibility

In some neighborhoods, especially after dark, open space provides a harbor for undesirable people and behavior; community gardens in out-of-the-way locations may suffer from severe vandalism for lack of guardians in nearby homes. Friendly, informed, watchful neighbors can make the difference between a safe community and a high-risk one. A policy of "eyes on the street"—or on the community garden—warns people that a piece of land belongs to the neighborhood and is not to be vandalized or used for inappropriate purposes.

SECURING A SITE

Going through the site selection process is a lot of work, but it could prevent the disappointment of developing a lovely community garden over several years only to find that it has been sold to a developer for new construction or that someone else is eyeing it for off-street parking.

Once you have decided on a site for your community garden, you will need to secure an agreement for its use. The most permanent option is to purchase the land, ensuring that all of the improvements put into the site will benefit your group for years to come. In most cases this will take money and legal advice. A less permanent and simpler option is to lease the land, thus eliminating the purchase price, often a substantial sum. To negotiate a lease agreement, try to get assistance from either a lawyer, a real estate agent, or someone familiar with real estate negotiations. This way you will better know how to approach the property owner. Many such professionals may be willing to donate advice for your project as a community service.

half an hour's work at a slack time can save you hours of back-breaking labor. If you don't have access to this type of equipment, round up as many friends and neighbors as possible to help you. Put the garbage in boxes or bags that can be handled by regular trash collectors. If possible, secure the use of a large dumpster through a local public works or housing rehabilitation agency, and hold a major clean-up to fill the dumpster.

Grade the site to make it as smooth and flat as possible. Depending on how level the site already is, this may be done by hand or may require a bulldozer or front-end loader. Again, if earth-moving equipment is available, it is worth the effort to get that help. Otherwise, pickaxes, shovels, wheelbarrows, and lots of people will come in handy. Either way, your goal is to eliminate any bumps and holes that will get in the way of your garden activities. If your site is on a slope, consider terracing to make flat planting surfaces in stairsteps. For this you will probably need the help of earth-moving equipment, as it involves cutting into the hillside at intervals and leveling the earth into stepped platforms.

Look at your site carefully. Examine the site to find out when and where it is shaded by the trees or buildings surrounding it. Make sure you know which side faces south. The sun's pattern across the site will be all-important when you are designing the plot layout. Look for natural features such as boulders, shrubs, patches of clover, or wildflowers that you may not want to disturb. Mark them so that everyone knows what to leave untouched.

An important first step in laying out your community garden is to sit down with the garden members and think out the group's preferences. Determine what you want to provide room for and how each relates to each—garden plots, seating area, compost bin, trash disposal, paths. Draw a plan of the garden (it needn't be fancy), and arrange the areas so that their relation to one another makes the most sense. For instance, you will want seating areas at a distance from the compost bin.

PATHS

Make sure your garden has several major paths for general use. These walkways should stretch the length of the garden and allow gardeners to reach their plots without walking across someone else's. Use stakes and string to lay the paths out neatly. Edges may be defined with bricks, border plantings, railroad ties or any other materials you may have at hand. The paths should be two to three feet wide, preferably covered with wood chips, straw, or clover. These ground coverings inhibit weeds and can eventually be turned into the garden soil. Pebbles or ground rock can also be used and will require even less maintenance.

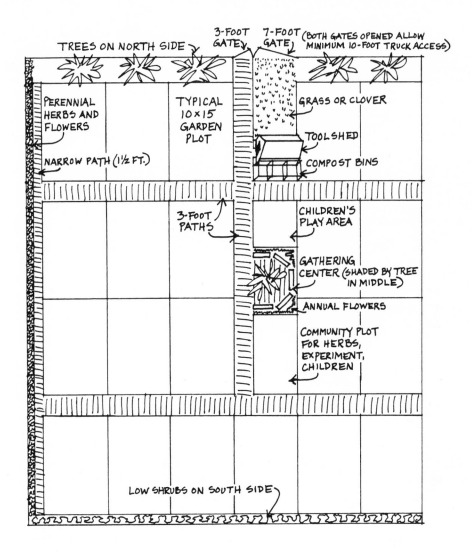

A sample site layout.

A GATHERING SPOT

Setting aside a place in the garden where gardeners can gather to sit and talk will encourage them to get to know each other. As gardeners exchange experiences and information they may even think up new ideas for improving the garden. A gathering place, furnished with tree trunks or any other kind of informal seating, will

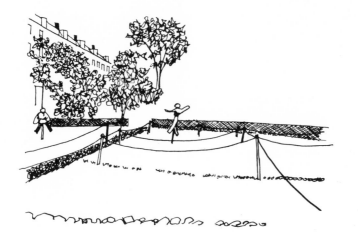

Plots and pathways
marked off with stakes
and string.

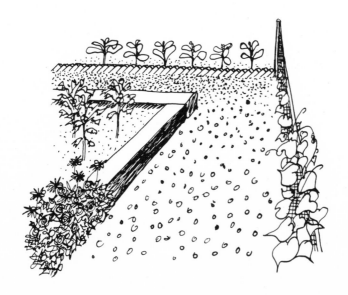

Railroad ties, bricks,
morning glories,
marigolds, and zinnias all
make good borders to
pathways.

have a positive effect on any garden. A shaded gathering place quickly attracts gardeners on hot summer days. A shady spot can be created by planting perennials such as fruit trees or by setting up a trellis and planting grape vines, which have additional edible benefits.

GARDEN PLOTS

Raised beds (see Chapter 13) make sense in an urban garden because they conserve soil and are ideal for space-saving intensive gardening. Raised beds also have clearly defined borders, helping to prevent accidental damage to plants. The borders can be made semi-permanent and attractive by landscaping or using railroad ties.

Plot sizes vary, but it is good to try to set a standard size for all plots—ten by fifteen, fifteen by fifteen, or twenty by twenty feet for example. Single gardeners can have one, large families several, if space permits. Lay out the plots with string, and scrounge or purchase boundaries. Three-foot stakes (or higher) at the corners of each plot will allow gardeners to enclose their gardens however they prefer. The more uniform the materials you use, however, the neater the garden will look.

THE COMMUNAL COMMUNITY GARDEN

Sooner or later, every community gardener comes around to the same frustrating limitation: lack of space. Despite all of the advances in intensive gardening and companion crops, often there just isn't enough space in the average community garden plot to support the quantity and variety of crops a gardener really wants to grow. In addition, concentrating a given selection of crops year after year in the same small area has the unpleasant side effects of depleting vital soil nutrients and perpetuating plant pests and diseases. After all, how do you practice crop rotation in a ten-by-fifteen-foot space? Somehow, we must find solutions to these problems.

Organizing the community garden as a communal venture is one way of addressing the problem of space limitation. This process can be approached in either of two ways.

The first method, and the easiest to implement for the average community garden group, amounts to simple division of labor and has the additional benefit of allowing individual gardeners to "specialize" in the crop or function in which each excels. At the beginning of the season, the group members should get together and determine which crops they wish to grow and how many plots should be devoted to each. Then, depending on individual skills and specialties, the group can decide who is going to tend which crops. If enough people are available, it is a good idea to assign one or more persons to functions such as common area weed control and central composting. All share in the harvest equally as each crop becomes ripe. Neighbors in the area who are unable to garden but are interested in produce could donate time or skills (canning, watering, etc.) in return for a share in the crop.

Obviously, this method will require much more coordination—and cooperation—than a one-plot-per-family garden; however, the results should be more than worthwhile. Soil depletion is solved by rotating plots where a given crop is grown on a yearly basis, and increased efficiency should permit a certain number of plots to lie fallow each year, sown to a "green manure" cover crop. Any legume, such as clover or vetch, can be seeded for green manure, so called because it helps fix soil nutrients. The only prerequisite of this system is that you have a group of gardeners willing to work with each other. If a crop does poorly, gardeners will remind their fellow gardener of his or her responsibility and, with each gardener growing a crop he enjoys and has success with, problems should be minimal.

A second communal gardening method, although more difficult to supervise, is more easily adaptable to subgroups within the garden. This method consists of two or more gardeners pooling their plots and planning, planting, and working them as a team. It should be noted that this method has been field tested with groups of up to four gardeners with moderate success.

One of the byproducts of a city environment is often a sense of isolation or lack of control outside the confines of the individual dwelling. Gardening as a community can be one of the fundamental steps for insular urban dwellers to begin to get to know their neighbors and to feel the basic group spirit necessary to community functioning. By getting people back in touch with the soil and with each other, community gardening fosters a rebirth of mutual respect between neighbors and restores to them a sense of control over their neighborhoods and their lives.

8

Soil

---◆•◆---

Linda Roth

---◆•◆---

Along with available water and sunlight, soil is one of your garden's most precious physical resources. Soil provides plants with support for their root systems and acts as a storehouse for water and nutrients. Proper preparation and care of the soil will be one of your garden group's most important—and most highly rewarded— tasks.

ASSESSING YOUR SOIL

The quality of the soil that comes with the site of your community garden will be determined by the history of the site. Has the site been farmed in the past? Was it ever built on? Was it leveled with "fill" at any time? The structure, fertility, water-holding or drainage capacity, and health of the soil will depend on the answers to these questions.

Unfortunately, most city soils have been rudely disturbed in the past, and in many cases there is no topsoil left at all. Soil mappers often label city soil as "made land," and in many cities that is no exaggeration. In Boston, for example, large areas of the city were created by leveling hills into the marshy bays of the harbor. In the process, topsoil went to the bottom, and subsoil and other fill wound up on the surface. The typical city site available for gardening is a vacant lot that has long been a dumping ground for construction rubble, unwanted subsoil, and household garbage. In addition, lots that have held houses have often been filled and leveled with demolition debris. Often what soil there is has supported crop after crop of weeds, with no replenishment of essential nutrients. What's left is shallow, dried-out, compacted soil that is unbalanced in texture, depleted of nutrients, and empty of vital organic matter. It may also be contaminated by heavy metals or other toxic substances.

Dig a small hole in your garden site and see for yourself. Check the depth; your plants will need a minimum of eight to ten inches, and ideally more, of penetrable soil. If you hit concrete or bedrock a few inches down, additional topsoil will certainly be required. How hard is it to dig? Heavily compacted earth could warrant a mechanical tilling job. Is it full of stones and bricks? Any large ones will have to be removed.

Now take a closer look at the soil itself. If it is dark and moist, soft to the touch, and alive with earthworms and other organisms, you are lucky; these are signs of a soil rich in organic matter. Organic matter, consisting of any plant or animal tissue, provides nutrients and moisture vital to plant development. It also supports beneficial animal life such as earthworms. If the earth has a dusty, whitish look and a dry, gritty feel, you should start thinking of sources of compost, rotted manure, or other organic additives. Extremes of soil texture are evident upon inspection. A sandy soil will feel rough between the fingers and will fall apart easily after being squeezed in the hand, whereas a soil with too much clay feels almost greasy, sticks together, and can be molded into shapes when moistened. Either extreme is undesirable, but both types can be improved by additions of organic matter. A sandy soil can be made to hold water and nutrients better; and a clay soil can be given increased drainage and aeration.

TESTING YOUR GARDEN SOIL

It is never too early or late in the season to take a soil sample as long as the ground is thawed enough to permit you to dig a six-inch-deep trowelful. While you wait

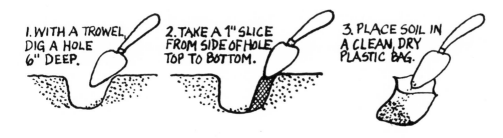

How to take a soil sample for testing.

for the soil to become workable in the spring, you can learn something about its condition by having it analyzed. Most county extension services across the country, which can be located through state agricultural departments or university agricultural programs, provide soil-testing services to the public in their area.

Before you reach for a trowel, though, you should be aware of what a soil test can—and cannot—tell you. Most testing centers analyze the chemical condition of the soil, its acidity, its nutrient status, and sometimes the presence of heavy metals or other contaminants. A fee is usually charged for the service, so it is useful to know what information these tests can provide.

A *pH test* offers the single most useful indicator of the chemical nature of a soil. The pH is measured on a scale from 0 (extremely acidic) to 14 (extremely alkaline), 7 representing neutrality. Most garden vegetables fare best in near-neutral soils of about pH 6.5.

Although pH does not tell directly the level of any particular plant nutrient in the soil, it does indicate the probable availability of most nutrients, to the extent that they are present to begin with. At the extremes of pH some nutrients become so unavailable that most plants starve, whereas other substances become readily available in toxic quantities. A pH test should be done every couple of years because most soils acidify gradually with time.

Your test results will include a pH reading and, if your soil is either too acid or alkaline, a recommendation for the application of either ground limestone or sulfur. Ground limestone will raise soil pH, and sulfur will lower it. These applications are usually given per 1,000 square feet, so you'll have to calculate the correct amount for your plot size.

Nutrient tests show you which specific minerals are plentiful or lacking in your soil. The test results list either the total or the available amounts of various elements required by plants. The difference between total and available amounts of a nutrient is that some elements in some forms break down too slowly for uptake by plant roots and are therefore unavailable, although they are present in the soil. Of all the elements, three—nitrogen (N), phosphorous (P) and potassium (K)—are most important to plant growth and are most commonly tested for. Nutrient test results may be accompanied by a recommendation for a standard chemical fertilizer (5-10-10 or some other balance of N-P-K percentages, in that order) to be applied at a given rate to correct deficiencies.

More comprehensive nutrient tests might analyze calcium, magnesium, and sulfur concentrations as well. Plants use these elements in substantial amounts, but they are less apt to be in short supply. Calcium and magnesium are present in most limestones and are added automatically to soil by liming to raise the pH; sulfur is readily available in most soils. Some analyses also report the levels of several trace elements required by plants in minute amounts. Although such information is interesting, it is difficult to use; trace element fertilizers can be costly, hard to obtain, and risky to use as most of these nutrients are toxic to plants at levels even slightly higher than what is normally in the soil.

No matter how exact your nutrient analysis may be, you should be aware that the recommendations you receive might not be as complete or specific as you would like. Most laboratories will not hand-tailor recommendations to put each nutrient in perfect proportion to the others in your soil. You may be given the standard advice: apply 4 pounds of 5-10-10 per 100 square feet. Also, recommendations are normally given in terms of synthetic chemical fertilizers, not bushels of horse manure or pounds of bone meal, so if you are using organic methods you may need to do some transposing. Exact fertilization requirements are more difficult to calculate in terms of natural fertilizer, but they are also less necessary. Serious nutrient imbalances are more likely to accompany the use of synthetic fertilizers, which are generally mixed in standardized proportions and do not include important trace nutrients.

Tests for contaminants may be either for your plants' benefit or for your own. Excess trace minerals, too much aluminum for example, can poison your plants, usually without posing any particular danger to you. Lead and cadmium won't hurt your crops, but they can be harmful to people—especially very young children—who eat plants grown in contaminated soil or breathe dust from soil carrying lead or other heavy (toxic) metals. Testing programs for heavy metals in soils are rare because the presence of these toxins in soil has only recently been recognized as an environmental problem. It is possible to locate heavy metals testing programs or even, as we did in Boston, to start one if necessary. Call your state or local department of public health, or university agricultural department, to see if there is a heavy metals testing program in your area.

LEAD AND YOUR VEGETABLE GARDEN

Lead is present in our environment from many sources. Automobile exhausts put lead in the air. Lead pipes leave traces of lead in the water. And where buildings painted with lead-based paint were demolished or where fruit trees treated with pesticides containing lead arsenate once stood, lead is in the soil. If you have small children or are pregnant, lead in your garden from the soil or the air could be a serious problem. Although lead is only one source of pollution in an environment in which new substances are found to be potentially dangerous almost every day, the presence of lead is a threat significant enough to warrant attention. As gardeners, we should be aware of the dangers of lead and take positive action to solve the problem of lead contamination.

Children under the age of six are affected most by lead because of their small size and early stage of development. Lead harms the kidneys, brain, and other body systems. Even children with continuous low levels of lead who show no apparent symptoms may suffer irreversible damage later. Pregnant women and nursing mothers should also be aware that the developing child is extremely susceptible to lead damage.

It is now known that the amount of lead in the soil directly corresponds to the amount of lead in the plants that you grow. Leafy greens such as lettuce, spinach, and chard and roots such as beets, turnips, and carrots accumulate lead the most. In comparison, fruiting crops such as tomatoes, peppers, corn, squash, cucumbers, and peas pick up very little lead. To date, scientific data on broccoli and cauliflower is incomplete. It is believed that a neutral pH, as opposed to an alkaline or acidic pH, may help reduce plant lead intake best if the soil has only a low level of lead to begin with.

You have reason to suspect lead contamination in your garden if the garden site

- had houses, factories, or other buildings standing on or near it
- is near a heavily trafficked road
- is near, or under, a painted bridge
- was once a landfill
- was once an orchard

Most often lead builds up from many sources, and a single cause cannot necessarily be pinpointed. Lead-based paint was used freely up to 1950, and any object or building on or near the site that was painted before this date may have lead particles left on it. These particles are often as fine as dust and can travel through the air and accumulate to significant levels even from sources that are no longer obvious. Lead, as a basic element, does not break down or leach out of the soil over time. On a garden site this means that, if lead was once present, it probably still is.

Be sure to protect children from lead by doing the following:

1. Test all children under six years old every year for lead whether they appear healthy or not. Doctors believe that lead can cause serious damage even before a child looks or feels ill. Look for your local public health service or lead testing programs. Many such programs are free or inexpensive.

2. Build a sandbox so the children do not play in the soil, as they will inevitably put their fingers in their mouths.

3. Have an area above the ground for children to snack at—a table, old wooden box, etc.—to keep food off the soil.

4. Always wash children's hands before they eat.

What can you do about lead in the air?

1. Select a site for your garden that is at least fifty feet away from busy streets or as far as possible from heavy traffic.

2. Plant a border of shrubs or tall or climbing plants such as sunflowers and morning glories, or build a solid fence to block out auto exhaust.

3. Grow leafy greens such as lettuce and spinach as far from the street as possible because it is impossible to wash all lead from their soft porous surfaces.

4. Wash all produce thoroughly before eating, and discard the older, outer leaves of vegetables in the cabbage family.

What can you do about lead in the soil?

1. Test your soil for lead. Call your local county extension service or your state department of agriculture to find out where lead testing is available. If you suspect lead contamination in a particular area, such as next to the street, take several samples from this area and mix them together into one. Do the same for an area that you suspect is lead-free. By combining several samples, you eliminate the possibility of an unrepresentative reading. A reading from a single point in the garden could be misleading; for example, that point might happen to be the place where a lead battery was once emptied.

2. If soil contamination is suspected, give planting priority to fruiting crops.

The basic elements of good gardening practice where lead is suspected. From *Lead in the Soil: A Gardener's Handbook*. Boston: Suffolk County Extension Service.

3. Keep the pH as close to neutral (6–7) as possible, and add organic matter to enrich the soil.

4. Always mulch around plants and cover exposed soil. Mulching reduces the settling of fine lead particles from the air onto the soil, as well as being a good horticultural practice for keeping down weeds and conserving water.

Lead contamination of the soil is a problem that has only been discovered in recent years, so many states do not have programs to deal with it. This was the case in Massachusetts in 1977, when concerned gardeners and scientists began to work together to institute public lead-testing programs and systems of outreach to make information on lead contamination publicly available. For more information on this organizing process, see Chapter 19, "Problem Solving Through Coalition Building."

GETTING "NEW" SOIL

Once all your prodding, inspecting, and testing are completed, you'll know if you're fortunate enough to be able to cultivate the soil you have or if you'll need to bring in additional soil by truck. If the existing soil is in fairly good condition but short supply, you can make the best use of it by creating raised beds. Remove the soil from the pathways, and consolidate it only where you will plant crops.

As you haul loads of topsoil from one location to another it is easy to lose sight of the value of soil as a fast disappearing and irreplaceable resource. Every cubic yard of topsoil you buy was scraped from some spot on the earth's surface that now has a negligible chance of being farmed again. As farmland dwindles and this natural resource becomes ever dearer, it is important to treat topsoil with respect. Keep it fertile and busy and recycle it when possible. When one large garden site in Boston was lost to construction of new housing, the gardening group involved hired a dumptruck and a front-end loader and carted off about sixteen truckloads of the soil to eight or nine other community gardens.

Topsoil is an expensive commodity, but if you need to purchase it keep in mind a few tips from experienced buyers that can save you money as well as trouble. First, buy in bulk. Prices and service are normally better for large purchases. It makes sense for several garden groups to place an order together. Also, shop around. Topsoil suppliers are listed in the Yellow Pages. Prices vary from six to twenty dollars per cubic yard. Unscreened soil is often cheaper, and if that only means the soil contains some rocks and roots, it could be a bargain. Also, get to know the dealers and find one amenable to working with community groups, delivering in the city, making split deliveries, and allowing you to check the soil you are buying. Do ask about the soil composition, and find out where the soil came from. Most dealers will be honest about these things over the phone, but they should also be willing to send a soil sample for testing through the mail. With purchased soil the nutrient and pH tests are not critical because these factors can be adjusted later, but the presence of heavy metals or excess soluble salts (which could be a problem, for example, in topsoil cleared from highway borders when roads are widened) are of greater concern.

Topsoil is sold by the cubic yard. To calculate the amount of topsoil your garden needs, you must know the area of the garden and the depth of soil required. Two simple formulas for calculating the number of cubic yards needed are as follows:

> For a layer 9″ deep: length (in feet) × width (in feet) ÷ 36
> For a layer 12″ deep: length (in feet) × width (in feet) ÷ 27

Conversely, to calculate the area that a given amount of topsoil will cover note that, for a nine-inch-deep layer, each cubic yard will cover thirty-six square feet of area; for a twelve-inch-deep layer, each cubic yard will cover twenty-seven square feet.

Making the proper arrangements for a delivery of topsoil is more important than it may appear. Be sure to remove anything that you don't want covered with soil from the site. If the existing soil is compacted, it will help to have the surface tilled before adding the topsoil, to improve drainage. It is essential to provide access for the delivery truck and to have someone familiar with the garden site on hand during the delivery. One example of what can go wrong if these precautions are not taken took place in a Boston community garden when the driver of a truckload of topsoil was unable to find the gate in a newly erected fence. In attempting to dump the soil over the fence and into the garden, he toppled a section of the fence and spilled half the topsoil onto the street. Fortunately, the soil was recovered by a quick rally of volunteers and the kindness of a construction worker who lent assistance with a front-end loader.

Apart from showing the truck where to enter the garden, a gardener who is present when the load arrives can arrange to have the soil dumped centrally to reduce the additional hauling that will be necessary later. He or she may also be able to persuade the driver to move forward while dumping to help spread the load. Unless an obliging bulldozer operator can help out, all further spreading will be up to you, using whatever wheelbarrows, shovels, and rakes your group can muster.

SOIL FERTILITY

Whether your topsoil is native or imported, it is well worth paying generous attention to its fertility. The nutrients removed by crops and drainage or water run-off must somehow be replenished if the soil is to remain fertile for future planting. Plants usually feel deficits in nitrogen, phosphorous, and potassium first. These elements are easily obtained in commercial chemical preparations and can be applied to fit soil test recommendations. These fertilizers have their drawbacks, however. Production of commercial fertilizers consumes large amounts of energy and uses up limited natural mineral resources instead of recycling available organic refuse. The quickly available chemicals from these fertilizers can be wasteful because of leaching from soil, can cause chemical burning of plants from concentrations that are suddenly too high, and can contribute to water pollution as the fertilizer runs off in a heavy watering. Before buying a package of 10-10-10 fertilizer, consider what soil amendments might be available locally, such as manure from stables or poultry farms, blood or bone meal from meat-processing plants, ash from a neighbor's wood stove, or seaweed from a local beach. Make a well-balanced compost to add to the soil, and try to contribute as much of the needed nutrients as possible that way. Although the chemicals eventually entering the plants will be the same no matter what fertilizer you use, those in organic form will be provided in a steadier, more balanced fashion that will contribute to the fertile condition of the soil over a longer period of time.

9

Compost

————◆•◆————

Linda Roth

————◆•◆————

Whatever the source of plant nutrients in your soil, there is no substitute for organic matter. Organic matter—the decayed remains of plants and other living matter—should constitute about 5 percent of the volume of a good agricultural soil. The soil in most city lots contains far less than that. Well-rotted organic matter benefits soil in so many ways that a list of them sounds like the virtues of an elixir. To begin with, the addition of organic matter improves a soil's structure, or arrangement of particles. Good soil structure is vital to proper root penetration and facilitates the distribution of water and air to plant roots. Organic matter also feeds worms and other organisms that contribute to soil structure. It serves as a reservoir for nutrients, keeping essential minerals from washing out of the topsoil, and, at the same time, it holds on to some contaminants and so helps retard their uptake by plants. Organic matter acts as a buffer, protecting soil against the extremes of pH, and corrects the problems caused by sands and clays, the two extremes of soil texture.

The list goes on. But before you can add organic matter to the garden soil it must first rot or break down. This process makes the minerals available to the growing plants as nutrients; however, the process also uses nitrogen. So if organic matter is breaking down in your garden, it is actually drawing nitrogen from the soil that must feed your plants. This all adds up to the need to set aside a separate area for making compost.

As extraordinary as the qualities of compost are, the process of creating it is actually quite simple, relying on the simple rotting processes that occur in nature. Making compost is simply a matter of piling up organic refuse and letting it decompose. Microbes attack the organic material and break it down into simpler compounds, releasing minerals to be recycled back into the soil and making them available to plants. In composting, as in other agricultural practices, natural conditions are controlled to obtain a better product at a faster rate.

A successful compost pile must provide a good environment for the microbes doing the work and, at the same time, prevent any loss of nutrients from the pile. The microbes most suitable for composting need plenty of air and water, a balanced diet, and temperatures on the hot side, around 140° F. To feed the microbes, it is best to mix materials high in nitrogen (manure, seaweed, fresh grass clippings, or kitchen scraps) with low-nitrogen materials, which provide carbon (straw, leaves, or sawdust). Ideally, the ratio of high-nitrogen material to high-carbon material should be one to three. For moisture, the pile should be watered periodically until the contents glisten but do not seem waterlogged or compacted. For air, the pile should be separated or turned, preferably every third day, for the first couple of weeks. A proper temperature is attained by providing the above conditions as well as by making the initial pile of sufficient bulk, a minimum of four feet in diameter is suggested, to allow it to heat up. In addition to speeding decomposition, the high temperature will help kill disease organisms and weed seeds in the compost.

Compost piles built all at once from collected ingredients work best, although it may be more convenient to let people contribute kitchen vegetables and weedy material as they become available. To speed decomposition, chop or shred the material into pieces no longer than one or two inches. For this purpose, a shredder is a good cooperative investment among gardening groups. Ingredients may be layered to make it easier to see and to set up the ideal proportions in the compost pile, but the neat, sandwiched arrangement will quickly disappear as the pile is turned. It is good to combine a couple of shovelfuls of soil with the compost mixture, but very little is required to provide the desired microbes. As decay organisms will be

DRY
LEAVES

GREEN
MATERIAL

STRAW

Layering materials in the
compost bin.

plentiful enough without them, the addition of commercial "activator" mixtures is
not necessary. Be sure to avoid adding diseased plants or weeds that have gone to
seed to the pile; meat and fat scraps should also be prohibited, especially in urban
piles, because they attract rats and flies. Although a compost pile does not require
a container, most community garden groups build a retaining fence to contain the
compost, principally for aesthetic reasons.

To keep a compost pile from losing nutrients, shape the top of the pile like a
shallow dish to prevent rainwater from washing down the sides. Cover the pile
during heavy or extended rains. Place absorbent bedding such as hay beneath the
pile. Do not add lime, and do not allow the compost to dry out or turn it too often.

The process of decomposition will take from several weeks to several months,
depending on the circumstances. A three-bin system in which contents can be
turned easily from one bin to the next encourages rapid decomposition and keeps

2"x4" FRAMING LUMBER TREATED WITH WOOD PRESERVATIVE

WIRE STAPLES

2" SQUARE WIRE MESH

A three-bin wire-mesh compost system. After the first bin is filled, its contents are conveniently turned into the middle bin, then the last bin, as it breaks down. New material is always put in the first bin, finished compost taken from the last.

community compost neatly contained. In the end, the compost will have shrunk in volume to about a third of its original size, and the resulting concentration of nutrients will be higher. You'll recognize the finished compost by its dark brown color, crumbly consistency, and pleasant, earthy odor. When all composted materials have completely broken down, this stabilized "humus" is an excellent soil amendment.

The finished compost, as well as lime and other soil supplements, should be added to the soil in spring prior to planting but after the soil has had a chance to dry out to a depth of six to eight inches. Tilling a wet soil will ruin its structure, so it is important to make sure that the earth is properly dry before cultivation. A ball of soil squeezed in the hand should break up at a light touch, and soil should not stick to tools or glisten where a spade has made slices. Spread measured amounts of fertilizer and ground limestone evenly over the garden, and work them well into the soil to a depth of several inches. Compost need not be measured because there is no such thing as too much compost in a garden. The difference in the soil's appearance now that humus has been added will be evident by the soil's darker color and richer texture. Continued additions over the years will show in the crops you raise.

10

Water

———◆·◆———

Jean E. Morse

———◆·◆———

Water is a basic necessity for all gardening. Luckily it is provided free in many parts of the country although only to a certain extent and somewhat haphazardly, from a gardener's point of view. As community gardeners, we need to develop systems to catch free water, retain it, and, when natural water becomes scarce, find a way to deliver "domesticated" water.

Water needs vary a great deal depending on your garden site, region of the country, and the season. Usually spring is the rainy season, providing perfect conditions for seeds and small plants to germinate and become established. By timing your planting with the weather and layering mulch on top of your garden soil to hold in moisture, little additional water should be necessary in the spring. Seed or transplant on an overcast day or right before rain is expected in order to make best use of nature's plentiful water supply during this season. Summer is usually the season when water becomes scarce. This is when you will need to use the water conser-

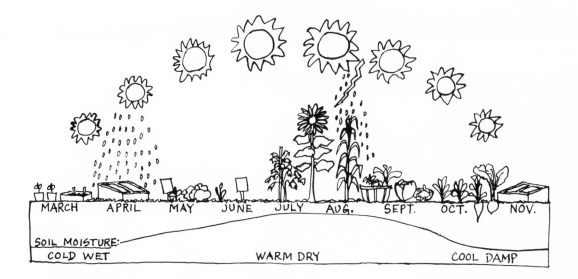

MARCH APRIL MAY JUNE JULY AUG. SEPT. OCT. NOV.

SOIL MOISTURE:

COLD WET   WARM DRY   COOL DAMP

The yearly cycle of rainfall. Drawing by Alice Evans.

vation techniques described in this chapter. By autumn the sun is weaker and the soil cooler; water evaporates from the soil and air more slowly and in smaller amounts. Plant growth is slower then, too, and therefore less watering will be required.

With this seasonal cycle in mind, become familiar with the options described here for water systems and water conservation, and apply them in the manner best suited to the needs of your plants, the site, and the season.

WATER CONSERVATION

Frequently the community gardener must deal with a limited water supply. For various reasons—problems of access, lack of convenient taps or hoses, difficulties scheduling water use, the cost of the water, the reluctance of abutters to permit constant use of taps on their property—gardeners will need to plan ahead, designing both individual plots and the garden as a whole with a sharp eye to water conservation.

The most important method for making the best possible use of the water you have is mulching. Mulch keeps soil cooler, retains moisture, and, in most cases, provides an irregular, porous surface, which increases dew condensation, thereby returning a portion of summer's high humidity to the plants. You can use a wide variety of materials as mulch—anything cheap, convenient, biologically safe, and reasonably portable will do nicely. Popular mulches include newspaper (no colored ink because of lead), hay, straw, leaves, and plastic sheeting. Optimally, the mulch layer should increase from an inch or so in the spring to up to four inches during the high summer and fall. Another incidental benefit of mulching is the accompanying decrease in weeds. Weeds simply can't get a foothold if light is prevented from reaching them. Mulch saves labor, eliminates weeds that are a harbor for pests and disease, and reduces competition from weeds for nutrients and water.

However, even the best mulch will not eliminate the need for a certain amount of additional watering during the drier months. The object is to get the most out of the least water. Spraying the ground with a garden hose takes a great deal of time and is about the least efficient method of watering a garden. Spraying permits a lot of water to evaporate, water that belongs in the soil, not in the air. Each gallon of water delivered to the soil directly will provide an inch of water for two row-feet, enough for a week, for most crops. By the same ratio, a ten-by-fifteen-foot plot, consisting of fourteen rows one foot apart and ten feet long, will require seventy gallons of water per week. These sample dimensions will be used in the various examples further on.

Plastic sheet mulch must be laid down before seedlings are planted. Slits are then made in the plastic to plant the seedlings.

WATERING SYSTEMS

The simplest source of water for a budding community garden is a neighbor's water supply (preferably an outdoor faucet), if you can get permission to use it.

For this you will need only a hose (or hoses). A more sophisticated water supply can be made using black poly vinyl (plastic) pipe, attached to the neighbor's meter directly or submetered. This type of system is durable and worth the five hundred dollars or so it will cost, including the submeter. Keeping barrels on the site can help to conserve rainwater or can help to minimize the inconvenience to the neighbor whose supply you are using if you fill them on a regular schedule to which everyone adheres. If barrels can be placed where they will catch the water coming from a downspout from a roof, you will have a large portion of your water needs provided for free. This source will go a long way in the hot, dry summer months, when rainfall tends to come in short, heavy downpours.

One of the first items the gardener should have is a simple rain gauge. This can be made out of any durable, straight-sided container marked (on the inside if possible) in half-inch increments with an indelible pencil or waterproof marker. A one-quart bleach bottle with the neck cut off makes a good rain gauge. Set this in the middle of the garden in an unobstructed spot, preferably sunk an inch or so into the ground for stability. Check after each rain or watering to determine the exact amount of precipitation you have received and how much more, if any, you need. This will enable you to make maximum use of whichever of the following passive irrigation systems you choose to employ.

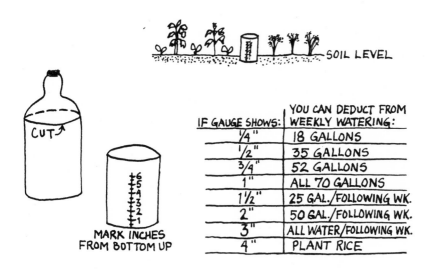

IF GAUGE SHOWS:	YOU CAN DEDUCT FROM WEEKLY WATERING:
1/4 "	18 GALLONS
1/2 "	35 GALLONS
3/4 "	52 GALLONS
1 "	ALL 70 GALLONS
1 1/2 "	25 GAL./FOLLOWING WK.
2 "	50 GAL./FOLLOWING WK.
3 "	ALL WATER/FOLLOWING WK.
4 "	PLANT RICE

CUT

MARK INCHES FROM BOTTOM UP

A simple rain gauge can help determine how much water your garden needs.

Remember, the only thing "passive" in this process is the water. Getting it to the garden will take plenty of activity. Water weighs 8 pounds a gallon, making a five-gallon bucket 40 pounds. Seventy gallons a week adds up to 560 pounds; this means you will need to make either fourteen trips with a full five-gallon bucket or twenty-eight trips with one half full—only 20 pounds. Bear all this in mind and do not strain yourself. That water does not have to be brought all at once, and with luck and a good rain gauge some or all of it can be eliminated.

A jug irrigation system is one of the cheapest and simplest watering methods. All that is required is a flock of gallon jugs (milk, juice, and spring water containers are ideal), a needle, a cork, and patience. A good funnel also helps. Stick the needle in the cork and heat it on the stove. Then burn five or six small holes in two opposite sides of each jug. Burn a couple of holes in the top of the handle as well for venting. Bury the jugs about six inches deep between rows, packing the dirt around them firmly. Set one every two feet; for a ten-by-fifteen-foot plot, you will need sixty-five. This will neglect the two end rows a bit. Compensate by planting

Jug irrigation.

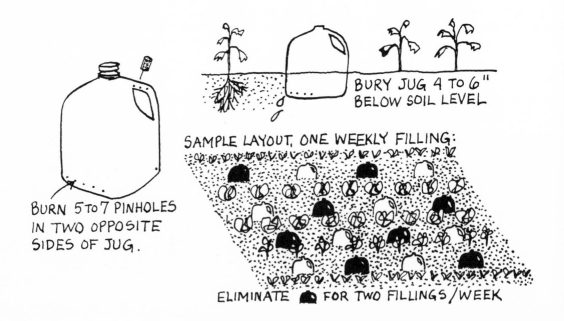

BURN 5 TO 7 PINHOLES IN TWO OPPOSITE SIDES OF JUG.

BURY JUG 4 TO 6" BELOW SOIL LEVEL

SAMPLE LAYOUT, ONE WEEKLY FILLING:

ELIMINATE ⬤ FOR TWO FILLINGS/WEEK

crops there that have lower water requirements. (Try marigolds—they are very useful as pest repellants as well as pretty.) One weekly filling of these jugs will provide all the water your plants require. Liquid fertilizer can be added to the water, as desired, for subsoil feeding. You can also use half as many jugs, twice as far apart, and fill them twice a week—it is up to you. Keep the jugs capped between fillings to reduce evaporation.

A second method of irrigation requires a little more ingenuity, along with a scavenged fifty-five–gallon drum and an old garden hose. This method can serve from one to four gardens simultaneously. Using an auger, bore holes in the sides of the barrel at the bottom. These holes should be slightly smaller than the diameter of the hose(s), of which you will need one for each garden. Seal the hose(s) in place with epoxy glue. Take the top off the barrel with a saw, if necessary. Save it, you may want it for a cover. Set the barrel at the head of your garden(s) and run the hose(s) in an **S**-pattern through the bed(s). Use an awl or icepick to pierce each hose (carefully) in appropriate spots, further apart near the barrel and more closely spaced near the plants. Double the end of the hose and clamp it. Then fill the barrel. For each ten-by-fifteen-foot plot you will need one and one-half barrels of water per week. If this is done cooperatively, the process can be relatively painless. But watch out for slackards: a two- or four-way hook-up automatically waters theirs as well as yours, whether they help with the watering or not.

Barrel and hose irrigation system.

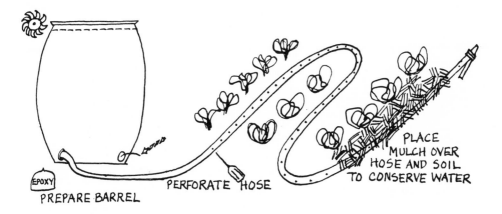

EPOXY
PREPARE BARREL

PERFORATE HOSE

PLACE MULCH OVER HOSE AND SOIL TO CONSERVE WATER

A wise person whose name has been forgotten once said that humankind owes its existence to a six-inch layer of topsoil and the fact that it rains. Certainly no human need is more basic than food, and nothing more necessary to growing food than water. No water means no garden. Soil can be improved or amended; sunlight can be simulated. Plant varieties can be interchanged and adapted. But there is no substitute for water. Make the most of natural water supplies, use them sparingly, and you will be rewarded with a flourishing garden, lower maintenance costs, and a sense of satisfaction in preserving an irreplaceable natural resource.

11

Fencing

Patricia Grady and Susan Naimark

Before I built a wall I'd ask to know
What I was walling in or walling out,
And to whom I was like to give offense.
Something there is that doesn't love a wall,
That wants it down.

Robert Frost, "Mending Wall"

These lines may be romantic and somewhat distant from the realities of modern-day community gardening, but they can serve to remind us of basic questions to be asked when considering fencing for a community garden. Fences help to define your garden space, keep animals out, discourage trespassers, and prevent dumping. Fences provide support for naturally decorative vines, climbing flowers, and even some of your vining vegetables. Although they can be expensive to install, any community gardening group with a few basic skills, muscle, and make-do spirit can build a fence. The fencing your garden group decides to erect will depend on the time, skills, materials, and money available. It will also depend on the particular conditions of your site, which will, in turn, determine the needs your fence must meet. These needs may include providing truck access to a specific area, protecting the garden from pollution from heavy traffic, keeping out trespassers, and letting in sunlight.

Once you determine the basic requirements for fencing your garden, you can look at the various alternatives. The different kinds of fences vary widely in price, durability, and ease of installation. Whatever fencing you decide to use, remember to leave a three-foot section for a gate for people, and a ten- to twelve-foot section for a gate for occasional truck deliveries of topsoil or manure. This section of the fence can be wired shut when not in use and does not need to be a gate full time.

TYPES OF FENCING

The following describes basic fences made of a variety of materials; nevertheless, these should not be viewed as the only alternatives. Creative use of available resources may lead to the construction of a type of fence you would never have planned on. The Dorr Street Gardeners in Boston, for example, discovered sites where the city was tearing down chain link fencing. They carted off sections to their garden and built their fence at practically no cost.

Wire mesh, commonly known as chicken or turkey wire, can be used to make the simplest fence a garden group can put up itself at low cost. Wire mesh comes in a variety of heights, gauges (thickness of the wire), mesh sizes, and roll lengths. The lower the gauge number and the smaller the mesh size, the stronger and thicker the fence. A 100-foot roll of galvanized welded wire mesh, four feet high, fourteen-gauge, with two-inch by one-inch mesh costs around eighty-five dollars from suppliers such as Sears. Vertical posts to support the mesh can be made of either steel or wood, both of which can be scrounged up or purchased new. The mesh can be attached to the posts with wire or, in the case of wooden posts, heavy-duty staples. The tools required for installing a wire mesh fence include a hammer, wire cutters, pliers, staple gun, and fence staples.

Although a wire mesh fence is not as strong or durable as either wood or chain link, it can be used to start a "living" fence of plants. Hardy climbing shrubs such as honeysuckle, wisteria, or the climbing varieties of roses can be planted alongside the fence. Over a period of several years these will grow up along—and replace the need for—your wire mesh fence.

Wooden fences can be purchased or scrounged. Picket or pallet fences are probably the easiest and best styles for wooden fences for do-it-yourself community gardeners. Salvage unpainted scrap lumber (avoid painted boards as the paint may be lead-based) and wooden pallets used for loading and unloading shipments of goods. Glass companies are excellent sources of wood because plate glass comes packed in wooden frames that can be used whole or taken apart. Local factories,

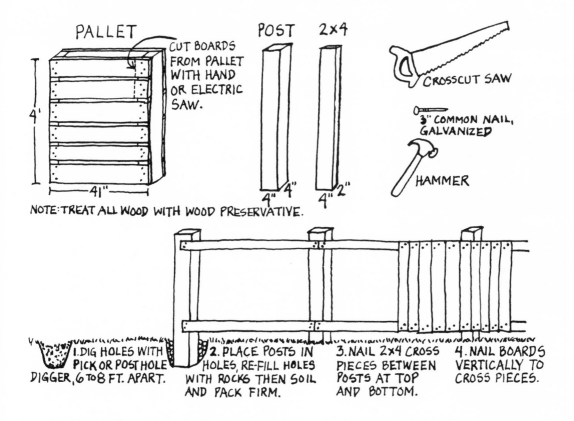

Basic fence design using recycled wooden pallets (devised by Ellen R. Sasahara).

distributors, and similar businesses often discard wooden pallets. These are made of oak, a dense wood that will last for years but is also difficult to take apart and nail together because of its density: oak may require the pre-drilling of all nail holes. If you let a local lumberyard know that you are a nonprofit organization, they may sell you wood at cost or donate scrap wood or four-by-eights for fence posts. Building demolition companies may also be able to supply unpainted wood; and recycled building materials can often be purchased at a fraction of the cost of new materials (look in the Yellow Pages under "Building Materials, Used"). Scrounged materials require transportation and are found only through persistence, but if someone in your garden group has the time the savings can be great.

Apply a wood preservative to all of the wood to be used for the fence to prevent or retard decay. Apply two coats of preservative at points where the fence will be underground or where it will touch the ground; apply an extra coat where two pieces of wood fit tightly together and to surfaces where moisture collects.

Preservatives safe for vegetable garden fences are copper chromium arsenate (arsenate) and copper naphthenate (Cuprinol). For information on the possible toxicity of wood preservatives, call your regional office of the U.S. Environmental Protection Agency and ask for the Pesticides or Toxic Chemicals Program.

Chain link fencing is probably the strongest and most expensive (two or three dollars per linear foot) type of fence you may consider. However, if you have the money to install a chain link fence, it should outlast wire mesh or wood fences by many years.

Although usually made of galvanized steel, chain link fencing also comes in aluminum-coated steel, aluminum alloy, and plastic-coated steel. It is available in heights of three to nine feet or taller. The basic components of chain link fencing are wire mesh fabric, horizontal cross posts, vertical tension bars (which keep the fabric tight) and fence posts. The lower the gauge number, the thicker and stronger the fabric. The smaller the mesh, the stronger the fabric. Chain link must be installed with steel posts set in cement and therefore requires more specialized installation than wood or wire mesh. Tools necessary for installation include soldering equipment, an acetylene torch, a hammer, a wrench, pliers, and a post-hole digger.

Whatever type of fence you erect, installing a ten-foot-wide gate for trucks and one or more three-foot-wide gates for people is recommended. Junk dealers often carry used gates, but any creative arrangement will do, provided it keeps out animals. If more security is required, a well-fitted gate with a combination padlock may be necessary.

ASSISTANCE

Hiring a professional contractor is the easiest but most expensive way to build a fence. If your group has the funds, be sure to get estimates from at least three contractors before choosing one to do the work. Some contractors may be willing to reduce the cost of installation if the group offers to assist. Professional estimates can also be used to measure the dollar value of "sweat equity" if you are planning to erect the fence yourselves and want to raise the funds for materials from donations.

Many funding sources will provide matching grants equal to what you can raise from other sources. If you are providing free labor, the price of this labor can often be the match they request.

Some cities have neighborhood improvement or open-space management programs, which may help you. Check with the planning department, parks department, or urban renewal authority in your area. Check the existing ordinances to find out what the legal responsibilities of landowners are. In Boston, owners are required to clear, fence, and keep clean the land they own. Because the city owned many of the lots desired by Boston Urban Gardeners, we could call upon city agencies to live up to existing laws, practically eliminating the cost of site preparation.

Do-it-yourself gardeners may feel that they already have all of the resources for installing a fence within their group. Although some types of fencing can be erected with relative ease, it is a good idea to have at least one person help who has some carpentry skills. A basic knowledge of the tools necessary for the job will ensure that the fence is assembled to last. This skilled assistance may come from a gardener in your group or from a neighborhood carpenter or contractor who is willing to donate an afternoon to the project.

PLANTINGS

Bushes, hedges, or climbing flowers planted along the street side of the fence can beautify the garden and screen out lead exhaust from traffic. Even tall weeds, kept clipped to the height of the fence, make a good screen. At street intersections, be sure to keep the plantings lower than the height of the fence for motorist visibility. Fencing that is further from the street can serve as a trellis for climbing vegetables and flowers.

MAINTENANCE AND REPAIRS

The amount of maintenance a fence requires will vary with the type of fence and the kind of garden it surrounds. The simplest wire mesh fence may last for years, although it is more likely to need yearly repairs due to its light weight. Wire mesh is also the easiest to repair; simply replace the damaged wire mesh or posts with new material. Sections can also be wired to each other to "sew" up a break in an otherwise good section of the fence. A small supply of wire, extra wire mesh, and a pair of wirecutters and/or pliers should be kept close at hand for repairs.

Wooden picket fences tend to break; they suffer from children and animals trying to squeeze through, over, and under, from badly aimed tools being thrown over them, and from vehicles running into them. Replacing broken pickets should not be any more difficult than the original construction was, especially if extra wood of the same type is set aside for future repairs when the fence is first built. Rotten wood can be detected by its softness; if you can easily sink your fingernail into the wood it has probably started to rot. This is most likely to occur on surfaces in frequent contact with moisture: horizontal boards with wide surfaces where water can sit, wood that is on or under the ground. Rotten or broken wood should be replaced every year. This task is good for a spring or fall work day.

Chain link fence repairs are the most difficult and probably the most essential if you have gone to the trouble of putting up such a sturdy fence in the first place. Although chain link should need little or no attention from year to year, accidents do happen. The most common damage to chain link fences that we know of comes from vehicles running into them.

Bent posts can usually be straightened by using a long pipe. Insert the pipe into or over the bent post at the top. This provides leverage that should allow anyone with a little muscle to straighten the post back to its vertical position. If vertical posts are capped, the cap must be removed in order to insert the pipe. On larger posts, a torch may be helpful to heat the bent part and soften it slightly, allowing it then to be straightened.

To replace a broken post where the base is still good, a smaller pipe can be inserted into the base to extend the post back to its original height. The original post and the new section should be secured with galvanized bolts. This technique can also be used to extend the height of an existing chain link fence, by inserting smaller pipes into all posts, and stretching new fence fabric horizontally across the entire extension.

Fence fabric, too, can be replaced or tightened by gardeners with the proper tools and a little skill. The easiest way to mend a tear is to wire it together, lacing the wire to pull the existing fence fabric tight and criss-crossing the opening to fill it in. Pull the existing fabric until it is firm but not too tight, to avoid bending it. For more extensive repairs, you will need to purchase or rent a fence stretcher, flat metal tension bars, wire cutters, and tension clips. New or recycled fence fabric should be hung loosely on the post and horizontal rails. It should then be attached to the opposite end post and to a temporary tension bar, and then the fabric is stretched tight. Cut the fabric one diamond from the end post, and attach it securely with a tension bar and clips. This method can also be used to attach a new section to an existing section of fence. After stretching the fabric, lace the new and old edges together with wire. If these instructions seem complicated go to a fence construction supplier and ask for a more detailed set of directions.

Some community gardens may not need fencing, but those that do should keep their fences in good repair. It doesn't take more than one unwanted trespasser (either two- or four-legged) to ruin months of hard work. Decide in advance who will take the responsibility to make repairs when fence damage is discovered, as it will inevitably from time to time. The more quickly damage can be repaired, the less likely your garden will be disturbed. Although Robert Frost's words may be true, "Something there is that doesn't love a wall, that wants it down," there *are* reasons for walling in your community garden—and being sure it stays that way.

12

Landscape Planting

◆•◆

Jane Lueders

◆•◆

There is more to a garden than its plots. Apart from the pleasure derived from growing and harvesting vegetables, gardens provide us with the opportunity to enjoy plants for their many other benefits. Landscape planting can provide the community garden with privacy and protection, as well as serving many other practical purposes. Selecting the appropriate plants for landscaping a community garden, particularly in an urban environment, requires imagination and some homework. Plants can serve a variety of purposes when planted with consideration for important garden needs and functions. Some practical ideas include planting

- thorny shrubs to form a living fence to keep dogs and other uninvited guests out of the garden
- tall-growing plants to provide a shaded seating area
- climbing vines to block out the auto exhaust of a busy street

- clover to reclaim poor soil
- trees or shrubs to block wind and noise
- and, of course, herb-, fruit-, and nut-bearing plants.

The environment can impose severe limitations on a plant's chances for survival, especially in the city. But many plants have proven themselves capable of withstanding adverse conditions. Each of the above kinds of plants has a character and special qualities of its own. The ability to adapt to certain conditions may vary tremendously from plant to plant. To choose the plants best for your garden, first consider the various aspects of each one to determine their suitability to your needs and to your garden.

PLANT SELECTION

Landscaping with annuals—plants that live for only one season—is the simplest way to beautify a garden. Most annuals, because they only live through one year's cycle, will not grow very large or bear edible fruit. On the other hand, they will bloom profusely and immediately the first summer. Annuals can be used to define borders, reduce garden maintenance (they keep down weeds), and reduce soil erosion. Some flowering annuals that are hardy enough to require little maintenance themselves include African daisies, gloriosa daisies, California poppies, morning glories, marigolds, and zinnias.

Planting perennials (plants that will live for many years) requires more careful planning. Meet with your gardeners to find out how people feel about landscaping and designing the garden. If everyone agrees, you can set aside zones for perennial plants, herbs, berries, flowers, or fruiting trees, depending on what you decide. Bulbs that flower year after year can be a good way to begin landscaping. They are easier to move than most other perennials if your site layout changes. They also require little maintenance once planted. Arbors, benches, tool sheds, and other garden features should also be considered during the early stages of planning, even if they are not constructed for several years. Taking someone's plot away for the sake of a new addition can be traumatic or even impossible, so try to anticipate your needs as a group and plan ahead. Keep in mind some basic rules of thumb when choosing your perennial plantings.

First, know the plant's form, size, and growth rate. Take care to choose a plant whose ultimate size is appropriate to the function you wish it to serve. Anticipate

Daffodil bulbs planted along the garden's perimeter provide a colorful border in the spring and require minimal maintenance from year to year.

the change in form and size that will occur with growth, and be sure the plant you choose is appropriate to its location (or vice versa).

Plant low-growing trees or shrubs on the south side of your site and high ones to the north to avoid shading the garden. Also, be sure not to plant perennials of any kind (such as strawberries, asparagus, shrubs, berry bushes, or trees) in the middle of the garden the first year. Live with your garden layout, and find out whether or not you will need to adjust your design in the next several years to make room for new gardeners.

Know the special seasonal features of each plant you are considering. From spring and into winter, your landscaping can provide a wide array of colors, forms, and fragrances. Combine plants whose flowers, fruits, branching patterns, foliage, and other seasonal highlights complement each other to create interest in the garden year-round.

Know the root system particular to each plant; a plant's ability to survive urban conditions of soil compaction, salt spraying, and drought is often determined by its root pattern. Plants with shallow, horizontal-spreading roots tend to be more sensitive to prolonged drought and soil compaction than those with deep roots. Shallow-rooted plants, on the other hand, are relatively fast growing and easily transplanted. Care should be taken not to place plants with vigorous shallow root systems where they might compete with the garden vegetables for soil nutrients and water, but remember that plants with deep taproots are particularly difficult to transplant.

Consider the plant's ability to withstand the extremes of climate in your area. A plant's hardiness varies considerably from one region to the next and depends on the temperature, rainfall, and soil conditions. Severe weather conditions can damage even the hardiest species, but it is wise to avoid plants that are susceptible to winter injuries such as consistent bud damage (due to severe fluctuation in temperature), broken twigs and branches, and trunk damage (due to exposure to winter wind). Urban areas often suffer from extremes in soil moistness; parts of a city may be dry because precipitation runs off the pavement instead of soaking into the ground. Conversely, some areas may collect the run-off water. Don't try to fight nature by selecting plants that are not suited to your region or site.

Be aware of microclimates on your site that could affect plants located within them. Microclimates are areas within the garden that have distinctly different growing conditions from the rest of the garden due to various surrounding influences. An area next to a brick building may have a warmer microclimate than the rest of the site because a brick wall absorbs and holds heat. Another area may be damper than the rest of the garden because it is at the bottom of an incline. Look carefully over your entire site, and try to identify these microclimates.

SITE PREPARATION

Preparation of the site before landscaping is important to ensure the plants' survival. Soil quality is a major concern. The soil contains the microorganisms, plant nutrients, and water essential to plant life. Poor soil may result in improper soil acidity, inadequate drainage and aeration, poor moisture capacity, and low fertility. Modification of the soil may be necessary.

Inadequate drainage and aeration occurs when there is too much clay in the soil. Peat moss may be added to the soil to loosen it. Although sand may be used to increase the drainage capacity of soil, it must make up about half of the total soil volume to be effective. Sand is inorganic and therefore relatively less beneficial than peat moss. On sites where soil is very poor, it may be necessary to remove the existing soil to a depth of two to three feet and replace it with a good quality topsoil or a mixture consisting of three parts loam, two parts coarse sand, and one part peat moss. Compost and/or well-rotted manure may also be added to build up existing poor soil.

PLANTING

Once the soil is prepared, flowers, ground covers, and other annuals can be planted. Woody plants and other perennials are a larger investment and should be planted carefully according to instructions. Spring is generally the best planting time for all plants because there is sufficient natural moisture in the soil and perennial plants are at the beginning of their growth cycle for the year.

Plant material obtained from nursery stock is more successfully and efficiently transplanted than that found growing naturally in woods and fields. In nurseries a plant's roots are pruned every few years to confine the root system to a smaller area and to encourage the roots to branch and develop a greater number of small roots and rootlets than they would ordinarily. Trees grown out of doors send out fewer and longer roots and consequently undergo greater shock when transplanted. This should not discourage you from experimenting; you lose nothing but time if a shrub transplanted from the woods does not take in your garden. Remember to be sensitive to the environment from which you are taking such plants. There is nothing productive about disrupting one environment simply to decorate another.

Plants may be moved with the roots exposed or with a ball of soil surrounding the root system. Often burlap is wrapped around the root ball to protect it further. Trees with bare roots receive a greater shock when transplanted and reestablish themselves more slowly.

Spring and fall are the best seasons for planting trees and shrubs. Holes for the roots should be dug and the soil prepared before the plants are moved to the site. Once there, the plants should be kept watered and shaded, and the roots covered with soil, wet sphagnum moss, or burlap. Detailed planting directions can be found in any good do-it-yourself home or garden landscape book. If planting material is purchased, the nursery where the purchase is made should provide instructions for planting.

Once the plant is in place, make a four-inch mound of earth around the edge of the planting bed to prevent water from draining off. A two- to three-inch layer of

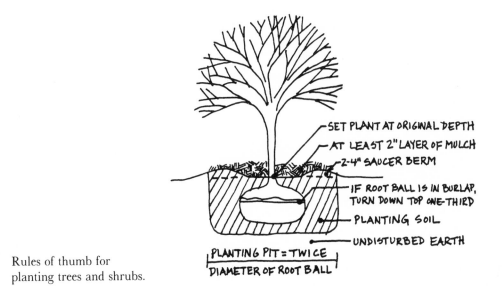

SET PLANT AT ORIGINAL DEPTH
AT LEAST 2" LAYER OF MULCH
2-4" SAUCER BERM
IF ROOT BALL IS IN BURLAP, TURN DOWN TOP ONE-THIRD
PLANTING SOIL
UNDISTURBED EARTH
PLANTING PIT = TWICE DIAMETER OF ROOT BALL

Rules of thumb for planting trees and shrubs.

mulch spread over the entire bed will conserve moisture, prevent wide fluctuation in soil temperature, and control weeds. Coarse peat moss, rotted sawdust, wood chips, or shredded bark may be used as mulch.

MAINTENANCE

The amount of watering a new plant needs depends on the type of soil it's planted in, the amount of rainfall it receives, and the size of the plant. In general, new plants should be watered once a week during the first year and more often during the drier summer months. Care should be taken not to overwater. Excess watering encourages the growth of root-decaying fungi and prevents water from traveling to the root zone. When you do water, however, be sure to soak the soil thoroughly to allow the water to reach the roots (two to two and a half feet for trees). Light watering will not meet the needs of the plant and encourages the growth of a shallow root system.

Trees and shrubs should not be fertilized for a year after planting, to allow the root system time to become established. Fertilizer should be applied either in the early spring after the ground has thawed or in late fall after the leaves have fallen and top growth has stopped. Summer fertilizing should be avoided. The application of fertilizer varies, depending on the soil, the type of fertilizer used, and the extent of the plant's feeder roots. The quick growth of a young tree will require a larger amount of fertilizer than an older tree, which requires nutrients only to maintain its health. Fertilizer should be applied to within a foot of a tree's trunk; applications made any closer to the tree may injure the trunk base and root collar.

Pruning affects the health and aesthetic appeal of a plant. Trees and shrubs that are properly pruned withstand severe environmental conditions better and require less additional effort to keep them healthy than those left unpruned. Remove dead or diseased wood, extraneous branches and twigs, low-hanging tree branches, and overly large limbs. Be careful to maintain the normal growth pattern and shape of the tree or shrub. Prune in the early spring to allow the plant an entire growing season to heal. Follow directions from your landscaping book, nursery, or landscape professional when pruning woody trees and shrubs. Pruning fruit trees is different from and more complicated than pruning other types of trees, so find someone with experience to help with this task.

Proper pruning and general maintenance, organized for each season from year to year, is well worth the time it takes. There is no quick substitute for years of care in creating the beauty of a well-landscaped community garden.

Part IV

TENDING
THE GARDEN

Photo by Read D. Brugger

13

The Plants

———◆·◆———

Susan Naimark

———◆·◆———

It's good to see who grows a nice garden ... to share ideas with people. Where I was growing up [in Florida] the farms were so far apart, someone down the road might grow potatoes well and you might grow good corn, but you never see each other. Here, you ask, "How do you do that?" and people tell you. You get to see each other. [The worst thing about gardening here is] wanting to give everybody some. You're proud of the garden and so you tell your friends. Then they all want to have some. You want to give to all of them.

—Bud Dupont, community gardener

Community gardening comes down to the same basic planting and harvesting involved in individual backyard gardening. The above quotation, though, illustrates that it *is* different in many respects. Hundreds of publications are available on basic gardening—in book stores, in your public library, and from the local U.S. Department of Agriculture county extension service. You should have no trouble finding a book suited to your interests, style, and region to serve as a basic reference. Before you select a book, however, it may be helpful to think about what planting considerations are unique to a community garden.

In a community garden you have a limited space that is yours to work. Many planting recommendations (including those on the backs of seed packets) are based on long, widely spaced rows, that do not fit the dimensions of a small plot. Instead, you will most likely be using intensive planting techniques such as intercropping, planting in raised beds, and succession planting (all explained later in this chapter), which produce high yields—enough to feed a family of six from a fifteen-by-fif-teen-foot plot, if well planned and tended.

Another consideration to keep in mind is that you will probably have neighboring gardeners on at least two sides. You need to consider the height of the plants you grow so that you are not shading the plot to your north. Planting corn may be a problem, and using stakes or a trellis for climbing beans or peas could produce significant shade as well. You may want to either eliminate tall plants completely or grow them in the middle of your plot, where they shade a part of your garden and nobody else's. You can then plant vegetables that do not require full sun behind (to the north of) these tall plants; some varieties of lettuce and spinach, for example, grow well in shady areas. Or perhaps you can team up with neighboring gardeners to grow tall crops cooperatively to work out shading problems.

Because the garden is not in your backyard, you should also consider realistically how often you expect to get to the garden and how much time you will be able to spend there. Take into account the amount of thinning, weeding, and tending various plants require. If you expect that you will have time only on weekends to work on your garden plot, consider the plants that are best suited to once-a-week tending and harvesting. In spring, most greens do well with little attention, although, as the weather gets hot, they bolt and become bitter very quickly. In the summer and early fall, root crops, Swiss chard, and peppers all grow slowly enough to go a week at a time without attention. On the other hand, summer squash can get too large and lose its sweetness in a week's time, and beans left unharvested get very tough.

One unfortunate consideration in many community gardens is the problem of vandalism and theft. Ask other gardeners what vegetables disappeared most often in previous seasons. The most commonly pilfered items in our experience are melons, pumpkins, and tomatoes. The first two vegetables require such a long growing season that taking the risk of losing any is hardly worth the investment unless the fruit is well hidden under foliage. Tomato plants, however, are usually so prolific that a few stolen fruit are not missed. In general, the vegetables that are the largest and most colorful—in a word, eyecatching—seem to be the most tempting to neighborhood thieves.

The presence of pests and diseases in a community garden must also be watched. If your neighbor's cabbage has aphids, there is a good possibility that they will find their way to your crops. The closeness of community plots increases the responsibility of each gardener to keep his or her crops healthy. Persistent pest problems vary from garden to garden and may even be different from the pests common to other backyard gardens nearby. One stubborn uninvited guest that we often find is the squash vine borer. This larva can be removed from the vine by hand but is nevertheless a problem throughout the community garden year after year. Remember that many insects or eggs winter over in dead plants or nearby weed patches. Destroying these affected plants (throw them in the trash, not the compost) can reduce next year's pest population. It is important to do all that you can to eliminate any pests and diseases as soon as they are discovered; still, there are some that you may have to learn to live with and work around. Always watch the application of pesticides or herbicides out of consideration for near neighbors or unknowing harvesters. Your garden pests may or may not be in your neighbor's garden, and anyone harvesting produce immediately after applications of pesticide needs to know to wash it thoroughly.

DECIDING WHAT TO GROW

What you grow in your garden is, in the end, a matter of personal taste, but here are some considerations that can help make your garden as economically and gastronomically (and nutritionally, see Table 6, page 172) worthwhile as possible.

- What is difficult to find in the market?
- What cannot be bought as fresh as it should be?
- What cannot be bought at reasonable prices?
- What provides a good return for the amount of growing space required?
- What does your family eat in large quantities?
- What is best suited to your region and the particular conditions of your site?
- What can be canned, frozen, or stored for winter use?

Once you have a shopping list of vegetables in mind, sit down with your gardening books and figure out the growing requirements of each. Consider the amount of space needed to grow the quantity of each item, any companion planting possibilities (plants on your list that grow well together), and the various planting dates and length of time to maturity for each plant (good planning could allow two crops in one space, an early crop followed by a later one).

With all of this information in mind—or written down to keep you from needing a computer to retain it all—draw a sketch of your garden plot. Do this to scale, one-half inch equalling one foot, for example, so that you can accurately mark off rows or beds for each item with a ruler. Indicate where south is on your plan with an arrow so that you know the direction of the sun, and mark any areas of special consideration, such as patches of heavier soil that have poor drainage, or shaded sections. Then it's a matter of solving a jigsaw puzzle, the pieces being the different types of vegetables you want to grow. Sketch out as many options for laying out your plot as you need until one plan seems to be most satisfactory. You may find you have to cut back the number of crops or the amount of each, or you may decide to grow more of something than you originally anticipated. There is no right or wrong solution; choose what will work best for you.

PLANTING TIPS

Raised Beds

Plant in beds rather than in rows to use space and topsoil most efficiently. Raised beds are created by moving soil from the pathways onto the bed where you are going to grow your plants. This way, available soil is not wasted on the walkways. The width of a raised bed should be determined by how far you can comfortably reach from either side—a width of three to four feet is standard. The length of the bed can be as long as your plot. As you add organic matter each year, the raised beds become higher, and you may need to add a retaining wall to hold the soil in

Creating a raised bed by hoeing topsoil from pathways onto the garden plot.

place. Bricks and scrap lumber are inexpensive and appropriate materials for this purpose. You will find that working on a raised surface is less backbreaking than ordinary ground-level gardening.

Spacing

Space individual plants so that at maturity they are almost touching. If a cabbage head will grow to twelve inches in diameter, plant small seedlings twelve inches apart. This intensive planting keeps weeds down and holds moisture in the soil because the ground is shaded by plants. Plants that are started from seed and require thinning can be scatter-seeded across the entire bed. Lettuce, spinach, and carrot thinnings can be eaten as they begin to crowd each other and need to be pulled up.

Succession Planting

Follow early crops such as lettuce and peas with later crops that can tolerate hot-weather planting and will mature by fall, such as beets and carrots. Or, plant the same crop in succession so that you have a supply all summer long. Beans are a favorite planted this way. Plant them every two weeks for a month in early summer for a continual harvest later on.

Intercropping

Plant quick-growing crops between slow growers, harvesting the first crop as the second begins to take over. Lettuce is a popular crop for planting between tomatoes and other slow-growing crops. Intercropping can also take into account the companionability of various plants. Pole beans can be interplanted among rows of corn so that the corn provides the staking for the beans to climb; basil interplanted among tomato plants not only repels pests but is said to enhance the flavor of the tomatoes; onions are a popular companion plant for many other vegetables because their strong smell is believed to repel pests.

Vertical Growing

Don't think of your garden as just a flat horizontal space. Use stakes and chicken wire to allow crops to grow upward, leaving the ground space for more planting. This method is used extensively by Chinese gardeners and produces an astonishing amount in a small space. Beans, peas, some squash, cucumbers, and even tomatoes will produce more per square foot if they are given vertical support than if they are allowed to sprawl on the ground. This method also eliminates the possibility of the fruit's rotting from constant contact with the moist soil.

SELECTING SEEDS AND SEEDLINGS

Send away for seed catalogs in January, and use them as a planning reference, even if you never order anything from them. Seed catalogs describe the different varieties of each available vegetable, planting schedules, which crops can be direct-seeded and which should be started indoors (or purchased) and transplanted, and the yields expected per plant or row. The descriptions are usually much more detailed in seed catalogs than on the backs of seed packets.

Seed catalogs are often the only source of information on varieties well suited to a small community garden plot. The seeds carried in hardware and dime stores are usually the commonest and most popular varieties, which are less specialized. In the seed catalogs, on the other hand, you can usually find dwarf or compact varieties (everything from cucumbers to carrots) which take up less space while maintaining high yields, varieties of cool weather crops (such as lettuce and cabbage) that are heat-resistant and will stand up to summer temperatures (this may be important if your garden is next to a brick building, which can actually make the soil next to it warmer), varieties that can withstand drought or water shortage, varieties developed to be resistant to specific pests or diseases that you know are a problem, and shorter season varieties that allow you to get more than one crop from the same space in one season. Using seeds from catalogs can also enable you to start earlier. By the time your local stores put seed packets on the stand, you can have your tomato transplants started in a sunny window or your peas in the ground.

Seedlings—plants that should be started indoors and later transplanted into your garden—can be started at home or purchased from a nursery. To grow seedlings at home, you will need a warm windowsill (60°F or warmer at night, 70°F or warmer in the daytime) that receives at least eight hours of sun each day. Seedlings can be started in any container with good drainage and two to three inches of soil. Garden packs, milk cartons with one side cut out, egg cartons, or paper cups all work well as seedling pots. If you don't have the time or space for starting your own seedlings, you can purchase them at hardware and dime stores, garden centers, and sometimes supermarkets. Any seedlings you buy should have strong, upright stems and no yellowing or brown leaves, which could indicate root damage. Sometimes commercial seedlings have not received the best care and can be stunted. You may be more satisfied by finding a neighbor or fellow gardener who will grow seedlings for you. In Boston some community gardens (one with a greenhouse) and one state school for the retarded grow excellent seedlings for sale to community and backyard gardeners. Proceeds go toward the gardens, and the seedlings are of the best quality. Using such local suppliers can provide you with varieties suited to community gardens that most garden stores do not carry. It is important to start with healthy seedlings if you want a healthy crop.

TOOLS

The basic tools you will need in any garden are a spade or fork, a rake or hoe, and a hose. Other desirable but optional equipment includes watering cans, hand forks, trowels, shovels, buckets, and a spray nozzle for your hose (see page 104). When buying any of these items, quality is worth the extra money; less expensive tools tend to break easily. The weight and size of the tools should fit your needs: the harder the soil, the heavier the tool needed to work it.

A wheelbarrow is very useful for hauling compost, weeds, rocks, and supplies. This is an appropriate item for community gardeners to buy together because of its cost, bulkiness, and relatively infrequent use. It may be stored at the community garden if there is a secure place for it. A rototiller may also be an appropriate group investment for the same reasons. However, because a rototiller is usually used only in the spring to turn the soil, renting one for a couple of days or a week once a year may be all your garden group will need. It is also possible to share other items, particularly hoses, heavier tools, and watering cans, if your fellow gardeners agree on a plan for their storage, security, and care.

All tools should be cleaned after being used and stored in a dry place. If properly taken care of they will last for years. When working with tools in the community garden, safety is very important. Don't leave tools lying where someone could step on them, and always keep sharp edges pointed toward the ground.

PLANTING PREPARATIONS

There are many methods for preparing your garden for spring planting. Different books may contain contradictory recommendations, which only shows that there is

Basic garden tools.

FORK
DIGGING & BREAKING UP THE SOIL; LIFTING POTATOES; SPREADING MULCH & MANURE.

HOE
LOOSENING THE SOIL; PLANTING; CHOPPING WEEDS.

SPADE
DEEP DIGGING; TURNING THE SOIL.

RAKE
LEVELLING THE SOIL; REMOVING DEBRIS.

Small hand tools.

TROWEL
FOR TRANSPLANTING

HAND CULTIVATOR
FOR WEEDING; LOOSENING SOIL AROUND PLANTS.

HAND FORK
FOR WEEDING; LOOSENING SOIL AROUND PLANTS.

no single correct method. Most gardeners, however, start by turning the soil, using a spade or shovel. This can be done with a rototiller if the area is too large or the soil too compacted for manual turning. Many community gardens will rent a rototiller and have the entire garden turned in one day. Some gardeners, however, prefer to turn their own plots by hand, especially if the plot contains perennials, such as mint or strawberries, that they don't want disturbed. If your soil is very thin or contains rocks or rubble, a rototiller may be damaged or may dig too deeply for good results. It is also difficult to rototill raised beds marked off by borders. Choose the method most appropriate to your site and your preferences. Setting aside a first spring work day for ground preparation can provide a time for gardeners to clean up winter remnants and help those gardeners, such as elderly people, who find it difficult to turn the soil.

After the soil is turned, compost and/or other soil amendments should be added. This replenishes the nutrients drawn out of the soil by last year's crops and adds organic matter. This can be a good time to have a truckload of well-rotted manure delivered to your garden for turning into each plot. After turning the manure into the soil, it should be raked to an even surface.

To mark off straight rows for planting, stretch string between two stakes and use this as your guide. You can draw a shallow line in the soil with a stick along the laid-out string, walking backwards. The seed is then dropped into this furrow and covered with soil. Tamp the soil firmly around the seed by walking heel-to-toe down the row or using the head of a hoe. Mark each row with the date, crop, and variety.

Many direct-seeded crops can be scatter-seeded across a bed. Beds planted this way should not be wider than you can comfortably reach from each side for weeding, thinning, and harvesting. Plants that can be scatter-seeded include lettuce, Chinese greens, spinach, carrots, mustard, turnips, and beets. Try not to spread the seed too thickly because this will make a lot of thinning necessary as it germinates. It is ideal to scatter the seed so that each falls about an inch from the next, although this takes practice. Thinning will need to be done several times, and later thinnings can be eaten.

Spring seedlings should not need much attention at first. Let germination occur based on rainfall and the normal soil moisture; this way there will be less risk that the seed will start to germinate and then dry out. Once you start watering a newly seeded row or bed, you must maintain that level of moisture or risk losing your crop.

When transplanting small seedlings, dig a hole slightly deeper than the depth of the transplant in its container. Some gardeners put a small amount of compost into this hole to give the seedling an extra boost of nutrients as it take root. Fill the hole with water and let it soak into the surrounding soil. Then carefully remove the seedling from its container and set it into the hole. Handle seedlings by the stems to avoid damage to growth. If seedlings are in peat pots they do not need to be removed from the container. Slits should be made in the peat pot (some gardeners prefer to carefully tear out the bottom of the pot) before planting to allow the roots to grow out. Peat pots should be placed in the ground so that the top edge of the pot is entirely below the soil level; the exposed edge will act like a wick, causing significant moisture loss.

Seedlings should be planted at the depth they were in the flat or pot they came in. The exception to this rule is tomatoes, which can be planted deeper to encourage new root growth from the bottom of the stem. Try not to transplant seedlings in the heat of the day. A cloudy day, the early morning, or evening are the best times to transplant. Don't be upset if the transplants remain limp for a few hours or a day; plants go into shock for a while when they are first transplanted. Transplants need to be watered frequently until they are well established, which can take up to a week.

Write planting dates and varieties on the map you sketched of your plot. This will avoid confusion if the markers in your garden are disturbed. It will also give you guidelines for next year if you follow through with notes on how each crop does. Some gardeners like to keep a journal in which they record what varieties they planted when and the amount each yielded.

AFTER PLANTING

When the soil begins to warm up and the spring rains become less frequent, mulch around your new plants with hay, straw, or any dry, nontoxic material. Mulch will hold moisture in the soil and keep weeds down. Warm, sunny weather means that weeds, as well as your crops, will grow quickly. More water and attention will be needed now than during cool, overcast weather.

Watch for damaging insects or diseases. If you suspect a problem and need help identifying it, consult a gardening book or an experienced fellow gardener. Treat any pests or diseases immediately so that they don't spread. Chemical pesticides are not recommended for use in community gardens because they spread easily through the water and the air to neighboring plots, where they may not be used. Remember that around 90 percent of all insects are either harmless or beneficial to gardens, so don't expect to keep your garden insect-free. Rejoice that your garden is an oasis for organisms that might ordinarily be absent entirely from many environments, especially in cities.

Many pests can be hand-picked from plants or treated using simple, nontoxic methods. It is also possible to introduce natural predators such as lady bugs and praying mantises for control of some pests. Consult an organic gardening guide for specific solutions to specific pest problems.

Your garden will need some form of fertilizer as the season progresses and plants draw available nutrients from the soil around their roots. Long-season crops such as tomatoes and melons tend to require more fertilization than quick-growing crops such as lettuce and spinach. Chemical fertilizers will deliver the needed nutrients to the plants quickly, but this high solubility also means that any nutrients not used right away by the plants will leach into the soil. Both well-rotted compost and manure tea, made by soaking manure in water to release its nutrients, provide nutrients in a form that will dissolve more slowly in the soil. This means that the nutrients are available over a longer period of time, which encourages steadier and more consistent plant growth.

As your plants mature and you begin to enjoy the harvest, all the planning, planting, and tending becomes worthwhile. In a community garden it is particularly important to harvest crops frequently. Ripe fruit that stands on the vine for any length of time invites theft as well as insect pests. If the harvest is more than you can consume, consider freezing or canning the surplus. There's no reason why a summer garden cannot be enjoyed right into the cold months of winter.

14

Keeping
the Garden Going

————◆•◆————

Martha Martin

————◆•◆————

After the garden is organized, built, and planted, the work is still far from finished. Even if people are working the soil, planting, weeding, watering, and harvesting conscientiously, needs will still arise from year to year. The garden coordinator(s) will need help with many of their tasks, and gardeners themselves may need assistance now and then. The needs of the garden and gardeners can be classified under the following categories: organization; physical resources; and continuing education and information.

ORGANIZATION

The garden coordinator is usually the person who "does it all" as far as the garden is concerned; he or she assigns plots, turns the water on and off, makes the rules, calls meetings, fixes the fence, and so on. Although this has been the usual pattern in our experience, it may not be the best. It is very difficult for one person to do everything, and others miss a lot of useful information and experience by not being more involved. The job of garden coordinator can be shared by appointing co-coordinators (two or more people with equal responsibility). Their role can be augmented by committees who volunteer to deal with specific issues or problems in the garden: watering, composting, maintaining security, working with children, and so forth.

Meetings seem to be a necessary evil in running the community garden. It is not easy to get people to come to meetings even at the best of times, so think about removing potential obstacles. Consider the safety and accessibility of the location of your meetings. Arrange rides for people without cars or for those who hesitate to go out in the evening. Or, hold meetings at the garden in the early evening during the summer when it is still light. Send a written notice (postcards are relatively inexpensive) at least a week before the meeting. Then, a day or two before the meeting, call people to remind them about it. If several people split up the list of gardeners, or you set up a phone chain, this task won't be so burdensome and can make all the difference in increasing attendance. Make the meeting a social occasion. Ask loyal gardeners or friends to bring refreshments. Consider making arrangements for child care, either at the meeting or at the homes of the individual gardeners who need it.

At the beginning of each meeting, the group should review the agenda and set goals for the discussion. It is a good idea to elicit everyone's opinions and ideas and to make sure that everyone goes away with a task to accomplish before the next meeting. Then, at that next meeting, the first item on the agenda will be to follow up on what was discussed last at the previous meeting. An important point to remember—especially in the beginning and perhaps throughout—is that someone will need to provide continuity from meeting to meeting. This person will most likely be the garden coordinator.

Even though you have a *community* garden, some people will be interested only in working their own plots. This is understandable but shortsighted. The garden as a whole has needs that go beyond the individual plots. There are common areas such as paths and borders, the fence that surrounds the garden, the water system everyone uses, and anything that may be unique to your garden such as a picnic area. These features will need maintenance.

Holding work days is a good way to get people together to work on the garden's common projects. These projects could include such jobs as weeding and mulching the paths and borders of the garden, building compost bins, repairing fences and water systems, planting trees and bulbs, or anything that the gardeners want to do to improve and maintain the common areas of the site. Spring and fall lend themselves to work days, marking the beginning and end of the growing season. However, if something needs doing, work days can be scheduled any time. Deciding ahead of time what work will be done on the work day will mean that supplies and tools can be collected beforehand to be ready when they are needed. Teams can take on specific tasks that day so that the maximum amount of work is accomplished.

Other kinds of events can help to keep gardeners and neighbors together and help to strengthen the feeling that the garden is an important part of community life. Potluck suppers or picnics, pictures in the local newspapers, and harvest festivals are some of the things garden groups have done to accomplish this goal. One Boston community garden hosts a "Wake up the Earth" festival every spring, attended by hundreds of neighbors who look forward to the event every year. Such gatherings give gardeners a chance to share ideas, to discuss issues, problems, and solutions, to trade information, and to see just how many other gardeners there are! The positive effects of such gatherings cannot be overestimated.

PHYSICAL RESOURCES

Because the need for physical resources will exist every year, the garden group will always need a means of obtaining these resources. The garden will need a continuing supply of topsoil and organic matter or fertilizer. Composting can provide one good source of organic material, but it may not yield the amount necessary for the entire garden. Donations of broken bags of peat moss and fertilizers can be solicited from local garden supply stores, or gardeners can stage fundraising events to purchase them collectively. Garden members can also chip in and purchase quantities in bulk in order to obtain lower prices for supplies. For those gardeners who have vehicles, a list of places that supply free manure and other materials will be extremely useful. For gardeners without a means of transporting supplies, it may be necessary to organize a collection to pay for transportation or to borrow municipal trucks or other vehicles. Once your group has had experience finding sources of free supplies, you may want to organize some of your scrounging into a once- or twice-yearly event.

CONTINUING EDUCATION AND INFORMATION

Gardeners will need continuing education and information—as they become more experienced, they will have more questions. A gardener who is interested only in

A yearly garden calendar.

the basics of planting during the first growing season is likely to be interested in much more the next year. Educational activities also serve to bring gardeners together.

Often the best teachers are the garden coordinators, who tend to be expert gardeners themselves, or older gardeners from farming backgrounds willing to share their past experiences. In some cases, however, people need more formal instruction, and it is important to collect information on where they can get agricultural extension publications, classes in horticulture, or material to read on their own. Demonstrations are also effective ways to teach and lend themselves to group activities. Building and maintaining a compost bin, cover cropping, and building raised beds can all be taught through demonstrations. Educational workshops can be built into a regular schedule if the interest is there. Monthly work days, from May through September, can be combined with one- or two-hour workshops on subjects pertinent to that particular time in the growing season.

After a few years, most community gardens begin to establish their own routines for these types of activities. Meetings, work days, and workshops become easier to organize as the organizers gain experience. People come to know what to expect, as well as what is expected of them, from season to season. As this occurs, you will come to understand—and truly enjoy—the benefits of community gardening.

15

Vandals
in the Garden

Martha Martin and Judith Wagner

The community garden established in 1973 by the Messinger Street Citizens Group is only one example among many of neighbors working together to better their environment. The group began not from an interest in gardening specifically but in gardening as an activity that would bring people of all ages outdoors to regain control of a local park that had been plagued by crime. After the community organized a twenty-four-hour park patrol, the garden was added to give the patrolers something to do on their shifts. It was used to draw neighbors out to the park on a continuing basis.

"My story is simple," coordinator Gareth Kincaid wrote to us.

In a word it is . . . survival! In 1973, when I moved into Mattapan, crime in our local streets was rampant. "What about a creative approach?" I thought to myself, hence I formed this citizen group. My philosophy was that other folks living here certainly

113

must feel as I do, that they would like to have a nice neighborhood. That's when I thought of our motto. "Be nice."

In our community our homes abutted a large piece of turf. It was a city park called Almont, which was getting a bad reputation for rape, mugging, drinking, and the list goes on. The neighborhood in 1973 was changing, and fast . . . blockbusting was in its prime. Families were moving out during the night; new ones moving in during the day. Things were happening so fast, no one really knew each other; no one was talking to their next door neighbor.

. . . I went to the local police station and explained what was happening and asked whether the police could patrol the area and give its citizens some protection. The reply was simply this: "We understand—but we do not have enough officers to just stand around and wait for an incident to happen." It was then that I realized that I must do something, and fast. Oh, did I pray! Then the thought came to me. If I could get the folks living here, both new neighbors and old, to walk our streets and control our park . . . then, and only then, maybe the good guys could outnumber the bad guys, and maybe they would move on. Hence, the Shangri-La Garden.

Vandalism is a general term that covers at least two types of problems for many community gardens: theft of produce and actual destruction of plants and property. The intensity of theft or destruction varies from year to year and from site to site depending on such things as the number and ages of children in the area, the location of the garden, the amount of unemployment in the area, school problems, and other social and economic factors. The most crucial factors in preventing vandalism, judging from the experience of garden groups we have talked to, are whether the garden is close to residents who can keep an eye on what is happening and whether people work in it frequently. If a garden is near people who care about it and work in it, it is less likely to be invaded.

Finding the garden you've worked on so hard ripped off or damaged is very discouraging, especially to a new garden group. Because you probably will not be able to prevent all the vandalism in your garden, it helps to think ahead about how to deal with the possibility of such damage occurring. There are many preventive measures you can take.

Prepare yourself and the gardeners for the fact that vandalism may occur. Talk about it early, and find out what ideas people have for preventing and repairing any damage. Suggest that people plant generously so that if some damage does occur, there will be replacement crops for people to share. Fix broken items immediately—holes in fences, broken pipes or posts, benches—to let vandals know you intend to keep at it. Try to develop simple designs for water systems, gates, and fences so that the results of vandalism are less serious and less costly.

People in the neighborhood can be your most important allies in combatting vandalism. Try to develop a good relationship with the young people in the area near the garden and, if possible, with their parents. If you have people with enough time and energy, it would be excellent to involve the young people with the garden, assigning them their own plots or work-day projects. Although this takes considerable effort, it can be a wonderful way to teach local children about gardening and invite them to be helpful in the neighborhood.

And, certainly get to know everyone in your garden. If you see an unfamiliar face, you will know you need to ask the stranger why he is there.

Even planting strategies can be used to prevent theft. Use unfamiliar varieties of plants to foil would-be thieves: white eggplant, yellow tomatoes, purple broccoli, purple beans. Bush crops (bush squash, beans, or cucumbers, for instance) often hide the vegetables better than vines and climbers. Some crops seem less attractive to curious children or vandals. Beans seem to be too much trouble; leaf lettuce and some less visible crops (like potatoes) may be left alone. But be prepared for the corn's irresistibility and the neighborhood kids' annual tomato fight.

Pick your crops at the earliest possible time so they won't tempt someone else. Even one messy or untended plot can attract "volunteer pickers," who may then turn to other plots for additional harvest. If you have to be away from your garden, have someone pick your produce in your absence. If necessary, plan to have someone present constantly in the garden by scheduling people to work their plots on a rotating timetable.

Should your garden be vandalized, replant right away whatever plants are ruined. Often you can get a late crop of near-normal harvest if you try again; this might cheer you up and will indicate to vandals that you intend to persevere.

Mr. Kincaid, organizer of the Shangri-La Community Garden, has developed a method for setting up block watches, an idea that might be applicable in many situations. Here are some of his suggestions:

1. *Stay alert.* Simply stated, know what is—and what should be—going on around you when you are in or near the garden.

2. *Window watching.* Gardeners who live within sight of the community garden should look out their windows off and on everyday and take notice of any strange vehicles or people nearby. This is not being nosy, it is being concerned.

3. *Telephone network.* Try to get at least one or two phone numbers of fellow gardeners and neighbors who live near the garden site. Then any concerns about suspicious activity can be quickly communicated among you.

4.   *Front and back door watch.* Gardeners who live near the community garden—or any neighbors to the garden—should from time to time open and stand in their doorways and look around. If they notice someone or something that seems strange or out of place, they should call other neighbors and have them go to their doorways to check it out. Often a stranger looking over the garden will leave quickly when people come to their doors or windows and start looking the stranger over.

5.   *Common sense.* Cool heads prevail: do not panic or let anger run away with you if you find signs of vandalism. Try to find out who may have done it. Children who play nearby or elderly people who are around during the day may have an idea. Approach the vandals, or their parents if they are children, and let them know that you know who they are. The Shangri-La gardeners have been known to put thieves to work weeding the garden, effectively discouraging them from repeating their offense.

Vandalism doesn't have to be inevitable. Your community garden can take the same active role in safeguarding the fruits of its labors as it has in making the garden a reality.

16

Growing
with Kids

—◆•◆—

Nina Gomez-Ibanez

—◆•◆—

When community gardeners complain of havoc wreaked on their plants, they often place the blame squarely on *"kids."* Community gardens are open spaces where adults work with their hands, enjoy themselves, and produce a variety of interesting looking plants, so it is little wonder that these gardens should fall prey to the curiosity and boredom of neighborhood children and teens. Rather than shaking their fists and erecting a taller fence, though, adult gardeners can present kids with simple and practical alternatives to pulling up carrots or throwing tomatoes. A garden is a place for people of all ages, if it is truly a community resource, and children can be encouraged to use it so that they develop an interest and respect for it that will carry into adulthood.

A PLACE FOR TOTS

Although some gardens are fortunately located near a school play yard or have received city assistance in financing a tot lot, most do not have these resources to entertain young children ready-made. With a bit of planning and a few hours of construction, a small space can be created that will help occupy children one to six years old while their parents work on the garden plots nearby. This space will also serve to protect young children if there is lead in the soil (see Chapter 8, "Soil"). As most parents know, small children can entertain themselves for an hour or so with very basic materials: sand, dirt, a bucket of water, a small blunt stick, pebbles.

With these magical ingredients in mind, gardeners can put together a sandbox frame with recycled lumber, railroad ties, old paving stones or concrete blocks. Most sandboxes are rectangular, but any shape will work as long as it holds one half to one ton of sand, which can be trucked in quite cheaply by a sand and gravel dealer. If the children's place has been set aside in the corner of the community garden, a triangular sandbox may be the simplest kind to build or a curved one might contrast nicely to garden plots laid out in straight lines. In any case, a sandbox is a good place to welcome the youngest gardeners, especially if effort is made on the part of the adults to keep it clean and appealing. A bucket or shallow tub of water is always a welcome sight, too, on a hot day, but it is not essential. Small gardening tools are a treat; children are naturally anxious to try out hoes and shovels. Lastly, something to sit and climb on can be created from several sections of logs or railroad ties set upright into the ground or from an old ladder cut in half and supported in a low bridge shape.

What about the "attractive menace" aspect of a children's place? Will older children throw sand and gang up on the little ones? Adults should make it clear to all children that this space is for tots only; rules are a necessary part of the community garden, and children understand this when they realize that they are participants, too, not outcasts.

ACTIVITIES TO INCLUDE THE
SIX-TO FOURTEEN-YEAR-OLDS

As they grow older, children who have been playing in or near the community garden will show signs of being ready to participate on a different level. Adults at the garden should be aware of the needs of this age group and set aside time to bring older children into the garden through activities that interest them. Depending on the age and inclination, older children can take responsibility for a variety of regular gardening chores such as watering, weeding, making signs, catching insects, collecting trash, opening and closing the cold frame, planting and transplanting in the spring, and seed collecting and harvesting in the fall. In fact, children in this age group are often ready to work on a small plot of their own (with varying levels of assistance). At the very least, these youngsters should be encouraged to participate in basic garden work alongside adults tending plots.

Bringing children into gardening activities requires a trusting and positive attitude on the part of the gardening adults and will pay off with returned trust and a sense of pride on the part of the children. Being excluded or kept away from the regular activities of the garden is hard on children and may be the beginning of a negative attitude toward the garden. Informally, kids can be asked (and challenged) to work with adults. "I bet you can't fill up a bag of trash as big as mine!" or, "I need help picking peas," or, "There's an extra watering can over there if you want to help me water." All of these are invitations and express interest and trust in the children. We gain a great source of help if we remember to communicate with young people throughout the year.

Formally, too, children in this age group can often be given group activities that stimulate interest in the garden and its care. Schools or garden groups can sponsor seed-starting demonstrations in the spring, providing the children with a soil mix of vermiculite and dampened peat moss and showing them how to seed a miniature windowsill coldframe in a milk carton. Cherry tomatoes and marigolds are usually a great hit with young gardeners because they are sturdy, productive, and hard to kill. Later, the children can make individual signs for each crop out of a stake or shingle with a picture of the vegetable glued on and shellacked. Clean-up days can be scheduled, with prizes awarded for the most work accomplished or the most weeds pulled. The prizes should always be related to the garden, however, emphasizing the idea that gardening has its own rewards. A package of interesting seeds, a hand cultivator, or a magnifying glass make good inexpensive prizes.

For a child who has already reached the point of boredom or antagonism by the time the garden is beginning, destruction may be his or her first impulse. Gardeners should try to find out who these children are, talk to them firmly—and their families if possible—but *always include them in some remediation job*. If antagonistic children pull up plants, have them help plant a new row. If they break a cold-frame cover, have them help you rebuild it. Redirecting the destructive impulse in this way usually brings about a positive or, at the very least, a neutral attitude toward the community garden.

WORK FOR OLDER TEENS

It is rare for a sixteen- or seventeen-year-old to take an active interest in the community garden. Whether because of peer pressure to spend time in sports activities, to work at a job, or just to "hang out," teens usually have better things to do, in their judgment, than spend time at the garden. If encouraged to tend a plot with a friend, or to make a garden in order to get in shape or learn a skill, a teenager will occasionally take on the project and succeed beyond anyone's expectations. Being able to work with younger children may also encourage an older adolescent to join in garden activities.

But more often, the young adult will have to be lured into working at the garden with the offer of a part-time job. Summer job programs for youth can sponsor such positions, but a gardener will often have to be on hand to supervise during work hours. Even if the city provides a supervisor, the gardeners must have a specific, organized work plan so that the youths will follow a sequence of accomplishments throughout the summer. Examples of jobs that have been done on sites in Boston's community gardens are path construction, border planting, tree and bulb planting, trenching for irrigation or water pipes, building a tool shed, tending a predefined area, constructing benches, and preparing soil for new plots.

Even without an official summer job program, a community garden group should consider generating funds for a part-time teenage worker. He or she could have a title—summer garden manager—and should be accountable for work on a daily basis. Local businesses concerned with the looks of the garden might want to sponsor such a position, or the garden group could hold a fundraising plant sale in the spring. The results are twofold: the garden gets a noticeable boost from the added energy and presence of these young adults; and they, in turn, get improved physical stamina, a set of new skills, and a growing respect for the existence of the community garden in their neighborhood.

By including young people in community gardens, you can grow much more than vegetables!

17

Solar Devices to Extend
the Growing Season

Susan Naimark

Even in cold regions of the country, it is possible to nurture gardens through most of the year by working hand in hand with that powerful source of life, the sun. Using simple materials and some ingenuity, the gardening season can be extended by a month at each end by the use of various devices that protect your plants from the cold. These range from something as simple as a sheet of heavy plastic laid over the growing area to a permanently enclosed structure such as a cold frame or greenhouse. These devices are called solar devices because they take advantage of the sun's heat by capturing it in an enclosed space where it builds up and is retained.

TYPES OF SOLAR DEVICES

The basic requirement for any solar device for growing plants is that it let sunlight in and keep cold air out. It may also have optional features such as ventilation to

121

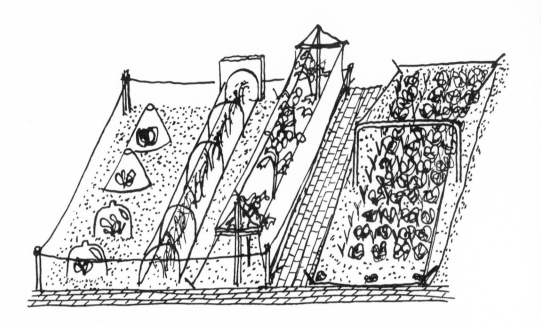

Cold-weather protection in the garden. *Left to right:* One-gallon jugs with bottoms cut out and hot caps of waxed paper or semirigid plastic can be placed over individual plants; plastic tents, sometimes called cloches (after the French bell-shaped plant covers of the nineteenth century), cover an entire row; sheet plastic can be used to cover an entire bed.

allow plants to breathe, insulation above or below the ground around the plants, heat storage in the form of rocks, an earth berm (dirt mounded up next to the solar device) or containers of water to retain heat, a night cover such as a straw mat or old blanket, and a wind screen. Solar devices may be made even more cold-proof by the addition of fresh manure layered under the soil bed to provide additional insulation and heat (from decomposition).

Any materials used for cold weather protection in your garden should be able to withstand large amounts of moisture. Rain, moisture in the soil, and condensation caused by temperature differences inside and outside all contribute to an environment high in moisture inside the protective device. Untreated wood may hold up for one or two seasons but will last many years more if painted with an oil-based paint or wood preservative. Safe preservatives for use around vegetable gardens are copper chromium arsenate (arsenate) and copper napthenate (Cuprinol).

A variety of translucent materials can be used, each with its own advantages and disadvantages. *Glass* is one of the least expensive (considering that it can last for

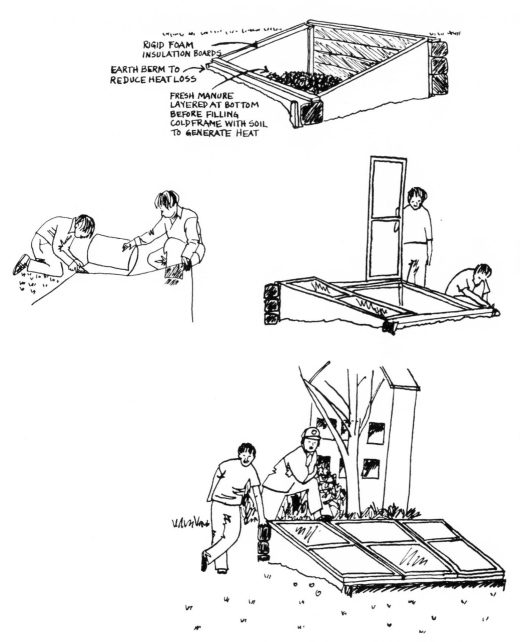

RIGID FOAM
INSULATION BOARDS

EARTH BERM TO
REDUCE HEAT LOSS

FRESH MANURE
LAYERED AT BOTTOM
BEFORE FILLING
COLDFRAME WITH SOIL
TO GENERATE HEAT

Top: Cold frame built with recycled railroad ties and plywood. *Middle:* Old wooden window frames are reglazed (glass replaced) with translucent fiberglass reinforced sheeting and installed. *Bottom:* The finished cold frame.

years) and can be recycled from old window sashes. It is also the most breakable glazing material. *Polyethylene* (the widely used plastic sheeting) is the cheapest and easiest to install but deteriorates in direct sunlight and does not hold heat in very well. *Fiberglass reinforced polyester* (a widely used variety is manufactured by *Kalwal*) is extremely durable but deteriorates in direct sunlight. *Plexiglas* or *acrylic* is virtually unbreakable but is expensive and also deteriorates in direct sunlight. The various plastics that deteriorate with exposure to the sun must be replaced every one to three years. If left longer, they will cloud and let less sunlight through each season. In the end, the glazing material that you use will depend on your situation: the likelihood of vandalism, how much money you have to spend, and how much time is available for yearly maintenance.

It is a big jump from these simple solar devices to a greenhouse, but each is essentially a protected environment that allows the sun in and keeps the wind out. The larger and more permanent nature of a greenhouse makes it more expensive and difficult to maintain. If you are interested in building a greenhouse, try using the simpler cold frame or other solar devices first to confirm your interest—and build up your skills—in this more specialized form of horticulture. Greenhouse plans and kits are available commercially, offering a wide range of sizes, prices, and degrees of sophistication.

USE OF SOLAR DEVICES

The purpose of using solar devices is generally twofold: to start plants early, before the beginning of the outdoor growing season; and to keep them growing longer at the end. Seeds germinate more quickly in warm soil, and transplanted seedlings grow faster and become established more quickly. How much you can push the season's start depends on the type of solar device you use. A tightly enclosed space may safely allow you to plant seeds or put seedlings out a full month earlier than usual. A loosely fitted tent may only be helpful on unusually cold nights around the normal beginning or end of the growing season.

Some vegetable crops are particularly hardy and can be maintained year-round in any region of the United States with the assistance of simple solar devices. These crops will not grow in the coldest months, when the light levels are at their lowest, but, if mature by that time, they will survive and can be harvested right through until spring. These vegetables include beets (for the greens only), Chinese cabbage, Chinese mustard, kale, collards, some varieties of lettuce, parsley, spinach, Swiss chard, and turnips (for the greens only). All are relatively fast growing and resistant to frost. Most seed catalogs will note the varieties that are frost-resistant, cold-resistant, or appropriate for fall or winter crops. These should be planted early enough in the fall to be mature by the time the cold weather sets in. An easy way to estimate planting time is to count backwards from your region's first frost date the number of days the seed packet gives until maturity for that variety. (If your frost date is October 15 and the packet of lettuce says forty five days to maturity, it should be seeded on or before September 1.)

Growing your own greens right through the winter is easier than most people think. At a community garden site, the winter growing space and maintenance can be shared, keeping the gardeners in contact with the soil and each other year-round. It can cut down on winter food bills and provide better quality produce than you'll find anywhere in the dead of winter.

Part V

BEYOND THE GARDEN: DEVELOPING LOCAL FOOD SYSTEMS

Photo by Janet M. Christensen

18

Land Ownership
Alternatives

◆◆◆

Patricia Grady

◆◆◆

One of the real things you have to increase and improve is ownership. . . . In the City of Boston less than 18 percent of the minorities own their own homes— or the feeling that the city is owned by them. They've done surveys and found that people in neighborhoods feel very estranged from the city government and have a greater identity with the neighborhood. One of the real problems you have is in maintenance of open space and having people feel that it belongs to them and not to the city. And you've seen a change when . . . there's been a community garden. It's not the city's garden, it's my garden, and if you're tearing down the fence or trampling the tomatoes or destroying the park, you're destroying my park because I worked on it and my kids are going to use it.
  —Isaac Graves, aide to U.S. Senator Paul Tsongas

Community control over open space is one important aspect of neighborhood improvement and stability. It is not uncommon for gardeners who have labored to transform a neighborhood eyesore (which usually exists primarily in low-to-middle-income urban neighborhoods) to find themselves losing their community garden sites to real estate speculators and developers.

Although the arrangement of permanent or long-term legal control of the property will not solve all the problems of a gardening group, it will allow gardeners to improve their site in ways that may not otherwise be possible or worthwhile. Garden yields can improve dramatically over the years with dedication to proper soil improvement techniques. Site stability can encourage this dedication and, in turn, encourage a more committed group of gardeners and lead to a better functioning organization. Acquiring ownership or long-term use of the land are ways

129

to ensure survival of the garden and can be a step toward community self-determination.

However, valid title to real estate cannot be transferred to an unincorporated association. For this reason, gardeners interested in buying or leasing their sites should form a nonprofit, tax-exempt corporation, often referred to as a community land trust. This process is often confusing, so a considerable portion of this chapter is devoted to the necessary steps in setting up a nonprofit corporation. Alternatives to outright purchase of property are presented later.

CORPORATE OWNERSHIP

A corporation is an association of individuals that the state government recognizes as a "legal person." It has the ability, called "powers" of the corporation, to hold title to land, make contracts, borrow money, and otherwise engage in business activities. The corporation's directors and officers manage the corporation in accordance with the procedures they spell out in the corporation's "bylaws," or rules by which the corporation functions. There are a number of advantages to being incorporated.

To begin with, an unincorporated group would find it nearly impossible to buy land because a deed to a piece of property held by an unincorporated association would be void. A garden club that is not formally incorporated can own land only through its individual members. This gives no recognition to the group and, conversely, puts all of the responsibilities of ownership on the individuals listed as the owners.

Furthermore, a corporation's existence as a legal entity is independent of changes in its membership. In other words, the membership could be turned over to a completely new group of individuals, but the corporation itself would remain unchanged. Individual members are not liable for judgments brought against the corporation in a lawsuit. Only corporation assets can be used to pay damages. For example, if somebody is injured at a community garden site that is owned by a corporation and sues the corporation as owner of the property, individual members of the corporation cannot be forced to pay any damages.

Once a group has been incorporated as a nonprofit organization in the state in which it operates, it may apply for federal tax exemption under IRS Code, Section 501(c)(3). Individuals who make donations to such groups can deduct the value of those donations from their personal income for tax purposes. Most foundations will award grant money *only* to groups that have this 501(c)(3) designation. IRS-recognized tax-exempt corporations are also exempt from state sales and income taxes (if the proper forms are filed), and may apply for local property tax exemption. (Some localities will grant this more readily than others.)

THE COMMUNITY LAND TRUST

Because the term *community land trust* is sometimes confused with a realty or business trust, definition of these terms is in order. A trust is created by a written document (trust agreement) that gives title to, and control of, real and personal property to a trustee for the benefit of a third party. The trustee can be one or more private individuals or a corporation. The third party is the beneficiary. The community land trust is a private, nonprofit, tax-exempt corporation that owns and manages real estate (land and/or buildings) for the benefit of the community.

The realty trust is often a for-profit (business) corporation set up to protect its individual members from tax liabilities. It is commonly used to purchase real estate, with members owning shares in the corporation.

The business trust is usually defined as an association of firms or corporations organized to control a particular industry. This type of organization often becomes the target of antitrust laws.

In some states "blind" trusts, set up to conceal actual ownership of property, are called land trusts. The community land trust is not to be confused with this kind of setup.

FORMING A NONPROFIT CORPORATION

Corporation laws vary from state to state. The state division of corporations, usually under the secretary of state's office, will provide information, assistance, and the proper documents for incorporation as a nonprofit, charitable organization. Specific statutes governing such corporations are in the state laws, copies of which may be found in business and law libraries. State bookstores often carry copies of the corporation laws.

In general, the organization must file, for a fee, a document called the "articles of organization," "certificate of incorporation," or "corporate charter" with the state division of corporations. Information required on this document includes the name and address of the corporation, its directors, its purposes and powers, its fiscal year, and the date of its annual meeting. The individuals who form the corporation initially, the incorporators, sign the articles of organization and submit them for the approval of the secretary of state.

The purposes and powers are crucial in determining both a corporation's non-profit status under state law and its federal tax-exempt status under IRS Code, Section 501(c)(3). To qualify for both, the corporation must be formed for any civic, educational, charitable, benevolent, or religious purpose. "Charitable" in this context generally means taking the burden of activity off government (for example, improving a poor neighborhood). The corporation must also be organized and operated for purposes beneficial to the public interest. These may include general educational purposes, reducing neighborhood tensions, or combatting community deterioration. Without emphasizing these charitable aspects of your group you may find it labeled by the IRS as a civic association 501(c) (4), which is tax exempt but is not able to receive tax-deductible donations.

The process of drafting articles of organization can be an important first step in a community group's assuming responsibility for itself as a legal entity. In the process of writing out your group's purposes, members of the group have the opportunity to share their individual hopes and plans for the organization. You may find yourselves spending many hours discussing ideas about community gardening, neighborhood land use, or other issues relevant to your organization. By formulating your group's purposes—keeping in mind the type of activities the Internal Revenue Service recognizes as nonprofit (such as educational and charitable activities)—your group can define its goals of long-term community service.

Just as critical for obtaining nonprofit and tax-exempt status as the stated purposes of your corporation is the statement of the powers of the corporation. This is simply a statement of the activities your group can perform to carry out its purposes. The community land trust has the *power* to buy land in order to carry out its *purposes* of promoting agricultural activities and preserving open space in the community. Chapters of the state laws dealing with corporations will list the appropriate powers that specifically need to be included.

Three requisites not always specified in state laws must be stated in the corporation's list of powers before it is eligible for federal tax-exempt status:

(1) No corporation member, trustee, or officer, or other private person shall benefit from or receive part of the corporation's net earnings. (2) No corporation shall carry on propaganda or otherwise attempt to influence legislation, or participate in any way in any political campaign on behalf of any candidate for public office. (3) If a corporation should dissolve, its assets must be distributed to a similar nonprofit tax-exempt organization.

The articles should also include a provision that the corporation will not carry on any other activities not permitted to 501(c)(3) organizations.

It is very helpful to get the assistance of an attorney experienced in real estate law for your incorporation process. This assistance should take the form of reviewing your articles of incorporation and bylaws. Someone in the group may know an attorney willing to donate his or her assistance. If not, call the local bar association and inquire about firms or individuals willing to give assistance free to community groups. This is a common practice in the legal profession and is called "pro bono" (literally, "for the good").

Once formally incorporated, an organization must file certain documents every year. For information on state requirements, contact your state division of corporations.

THE CORPORATE BYLAWS

In addition to the articles of organization, a community group seeking to incorporate must compose a set of bylaws by which to govern the corporation. The bylaws contain provisions regarding qualifications for directors and members, terms of office, meeting times and notices, voting procedures, and duties of the officers of the corporation. The bylaws may contain a further explanation of the purposes of the corporation, or any other matter not covered in the articles of organization. The bylaws are adopted by the directors (or members) at the first official corporate meeting.

It may be useful to review examples of the bylaws of other groups similar to yours. It is often tempting to "borrow" bylaws from other groups, but this should be done only with caution and much careful consideration. These bylaws will dictate the rules by which your group will operate and should fit your particular group's needs and style.

The corporate bylaws need not be approved by any governmental entity. There are few legal requirements concerning the contents of the bylaws. In general, this document is more flexible in form and may be more easily amended than the articles of organization. The bylaws may be changed by a simple vote of the members or directors. A change in the articles, on the other hand, must be voted by the corporation, then submitted to the secretary of state with a filing fee as an official amendment to the articles of organization.

APPLYING FOR FEDERAL TAX-EXEMPT STATUS

Within fifteen months of being organized, a nonprofit corporation should apply to the Internal Revenue Service for tax-exempt, tax-deductible status under IRS

Code, Section 501(c)(3). Once granted tax exemption, the corporation will be exempt from state sales and income taxes (but not without filing the proper forms) and is eligible for a nonprofit bulk rate mailing permit from the U.S. Postal Service. Most important, a 501(c)(3) exemption allows donors to your corporation to obtain a tax write-off for all contributions they make.

The application for tax-exempt status is standard for all states. The corporation must complete Form 1023 and submit it to the regional office of the IRS. The form asks for a complete explanation of the corporation's purposes, its funding sources, and its anticipated projects and expenditures. The IRS will also review the corporation's articles of organization to make sure that its purposes are consistent with the meaning of a public charity as defined by federal law. For more detailed information, see IRS Publication 557, available without charge from the nearest district IRS office. You may find the forms difficult to interpret and even overwhelming if you are not trained in legal jargon. You may find free assistance, however, from the district IRS office or from volunteer legal services, which are sometimes available to nonprofit organizations through local bar associations.

When the IRS notifies the corporation of its approval in the form of a determination letter, the corporation may accept tax-deductible gifts from foundations and charitable individuals. These gifts may include donations of property. In order to retain tax-exempt status, the corporation must file reports of its financial activities and programs with the IRS each year.

Although nonprofit charitable corporations may apply for property tax exemption, policy on granting such exemption to community groups varies from city to city. Land trusts in cities reluctant to grant exemption may instead apply to the assessor's office for a tax abatement. This is a reduction in tax, based on the profitability of the use of the property. A gardening group may receive an abatement based on its not-for-profit use of land as open space.

LONG-TERM LEASES

Some community gardening groups may find long-term lease of their sites preferable to ownership. The classic form of this kind of arrangement is a ninety-nine-year lease; however, leases can run from a few months to one, five, ten, or more years. Some states limit the duration.

Public institutions or businesses that own land in the community can be approached for long-term leases. Municipal parkland, if available, is another possibility. Vacant lots foreclosed by the city for delinquent taxes can often be leased on a year-to-year basis. Long-term leasing is often not feasible from a local government; most cities would rather sell the property to get it back on the tax rolls.

In negotiating a lease, keep in mind the following:

The landlord is the *lessor,* the tenant the *lessee.*

The lease remains in force even if the lessor sells the property, unless it includes a clause specifying that sale terminates the lease. For a long-term lease to be binding on a new owner, some states require that the lease or notice of the lease be recorded in the registry of deeds.

Some states require that leases running beyond a certain period of time be recorded in the registry of deeds. Others require long-term leases to be notarized.

Under the terms of a *gross lease,* the lessor (landlord) pays all expenses, including taxes, insurance, maintenance, and repairs out of the rent paid by the tenant. In the case of vacant land, however, the landlord will probably require the tenant to carry liability insurance. Under the terms of a *net lease,* the tenant pays some or all of these expenses, in addition to the rent.

A lease with the option to purchase gives the lessee (tenant) the right to buy the land at a specified price within a specified period of time. Usually the rent or a portion of it can be applied toward the purchase.

CONSERVATION HOLDINGS

State, county, or municipal conservation commissions can acquire public or private land for long-term or permanent preservation. Private conservation groups (which are sometimes well endowed) can also purchase land for permanent use as open space. In some cases these groups may be willing to arrange lease or use agreements with gardening groups. Community groups that are concerned about the permanence of open space but are unable to undertake ownership should contact their local conservation commission or any private conservation groups in their area to find out about this option.

Gardeners should review their options for securing land and investigate the relevant laws and agencies in their area. Land ownership may seem like a big responsibility, but remember that it is a *shared* responsibility in the case of a community garden group. Proper homework and legal assistance can allow even a small garden group to become land owners. The responsibilities of such an undertaking are significant and should be taken on in a serious and cautious manner. Equally significant is the fact that owning your community garden can result in stability and security for an important activity in the life of your community.

19

Problem Solving Through Coalition Building

———◆•◆———

Susan Redlich

———◆•◆———

In 1977, just as the idea of community gardening began to take hold, as community residents and organizers were beginning to see some heartening and dramatic results from their work, an unexpected problem threatened the very purpose of community gardens. The story of how this threat was discovered, handled, and continues to be monitored is an example of how group cooperation can go beyond the garden plots to solve major problems.

Curious and concerned about a garden that refused to grow even weeds, a member of Boston Urban Gardeners tested some soil samples for toxic elements. At the same time, two scientists—one from the University of Massachusetts and one from MIT—began a study of the pollution of soil by airborne lead. Also at the same time, another Boston gardener sent a soil sample to the regional U.S. Environmental Protection Agency (EPA) and requested help in determining its heavy metal content.

The results of the soil tests were, in two cases, alarming. The BUG member alerted another member of the Boston Urban Gardeners coalition who, in her role as director of a state agricultural program, arranged a meeting of concerned gardeners with the city officials sponsoring the contaminated garden. The university scientists were invited as resource people and, almost by chance, the EPA scientist learned of the meeting through a colleague. Early discussion determined that the problem with lead in the soils was serious enough to warrant the formation of the Boston Ad Hoc Task Force on Lead, which continued to meet through the Division of Land Use of the Massachusetts Department of Food and Agriculture. This task force became a way to organize existing agencies, individuals, and interest groups to deal with the problem of lead in the soil, a problem that had gone virtually unrecognized before 1977. The task force has been an experiment—and a successful one—in cooperative problem solving, a rare exercise involving a controversial public health problem.

The task force relied initially on the volunteer efforts of all its members. Meetings were called, strategies discussed, individuals and agencies contacted for assistance with various tasks. One advantage to this volunteer approach was that it was not held back by lack of funding or organizational endorsement. Another was that the mix of agency people who were involved, at first voluntarily, helped pave the way for the eventual contribution of temporary, then full-time staff. Community gardeners themselves donated their time to write proposals for funding, which led to a grant to pay for further educational outreach work.

The lack of official endorsement for a lead-poisoning intervention program meant that any assistance from the government depended upon the willingness of individual administrators to take on extra work unofficially. Gaps in policy and the scarcity of regulations made it difficult to pin the responsibility for clearing up the lead problem on any one organization. A fear of setting off negative public reaction with possible political repercussions contributed to the reluctance of some agencies to acknowledge the problem. Others were concerned that, by acknowledging the problem, they could be held accountable for solving it.

The response to the lead problem was stimulated by the work of individual urban gardeners concerned about the unknown hazards of their food production. Their persistence in seeking answers and their willingness to carry out the volunteer work of collecting samples, trucking topsoil, attending meetings, and getting information out to other gardeners made progress possible. In addition to concerned gardeners, the task force included professional scientists and government administrators who served in an advisory capacity. The participation of these professionals was important in convincing several government program and department heads that a problem existed and warranted their attention. The shared responsibility of the task force allowed the Massachusetts commissioner of agriculture to endorse participation by sympathetic staff of the Division of Land Use within his department, rather than canceling this participation for fear of accepting sole responsibility for a controversial problem. Regular communication from the task force to the director of the County Extension Service made him understand the need for a coalition that included the participation of his program.

Participation in several conferences and workshops that included the topic of lead poisoning or soil contamination brought encouragement and requests for information from groups in other major cities. This furthered the meaning of the task force's work as well as the motivation of various task force participants. By defining the issue within the broad context of environmental and public health, the base of knowledge and support was strengthened from many different angles.

From the start, the commitment to serving gardeners, despite the tentative and incomplete nature of the task force's recommendations, helped to educate gardeners, to prompt others to seek information, to develop a constituency for the task force's work, and, finally, to focus media attention on the issue, giving the group renewed stimulus and power.

The measure of success over the five years since the task force was formed is substantial. A lead-testing service has been created, run by a staff that handles public information, laboratory work, post-lab followup, and documentation. Over three thousand gardeners have sent soil samples in for testing and have received appropriate recommendations according to lead levels recorded. Workshops and news articles have covered the issue. The U.S. Department of Agriculture Extension Services now process the before and after lab work for soil samples in their jurisdiction. Scientists have teamed up to conduct a study of the rate of lead uptake by different vegetables. The task force used the results of this research to fashion a set of recommendations to gardeners for mitigating the effects of lead in the soil at various levels. Careful attention was paid to provide clear information that did not frighten or discourage gardeners but that offered practical and manageable solutions.

Resources that have been secured to support ongoing efforts in the antilead campaign include a provision for internship assistance, state and federal funds for lab work, and a grant from the National Center for Appropriate Technology to publish a citizen booklet on lead contamination. Prompted by a high lead reading from the task force's screening of gardens and adjacent playgrounds, the City of Boston removed contaminated playground soil and replaced it with sand. A demonstration grant was awarded to the State Department of Agriculture, Division of Land Use, to investigate the establishment of a compost facility that would recycle locally available organic waste back into garden soil that needed rebuilding and amending by the addition of uncontaminated material.

These successes could not have been achieved by one community garden group acting alone. The citywide and, to a certain extent, statewide activity of the task force has created a wedge with which to widen public attention and interest. By holding various public agencies accountable, through public pressure and use of sympathetic individuals within these agencies, resources were gathered with which to tackle a previously ignored public health problem. The coalition approach allowed individuals and agencies to participate, each with their unique skills which, when combined, helped resolve a complex and serious community issue.

20

Farmers' Markets

———◆•◆———

Greg Watson

———◆•◆———

The establishment of farmers' markets in cities and towns throughout the United States has demonstrated that city people and farmers alike have begun to take control of local food economies. The small farmers who market directly to the consumer can save over half of each dollar that is usually alloted to intermediaries, and the buyers get fresher and cheaper produce. The produce sold at a farmers' market is usually picked within twenty-four hours of sale. The buyers do not have to pay for the fancy packaging common to supermarket produce, and often there is more variety.

Farmers' markets depend on cooperation between urban and rural communities. Urban and rural dwellers have discovered that they can contribute to improving the quality of life in their respective communities by setting up a direct market outlet for local farmers during the growing season. A farmers' market may be one way for successful community gardeners to sell any surplus produce they have grown. It may also be a way to increase access to fresh produce for nongardening neighbors.

Boston, which is at the end of the line in national food distribution, never has more than a ten-day food supply at any one time. Many other cities and areas are similarly dependent upon a costly, energy-consuming food production and distribution system. In the Boston area, farmers' markets provide residents of eight neighborhoods with a low-cost, energy-conserving, ecological strategy for dealing with food prices that are among the highest in the nation. The markets, begun in 1978, have operated on a very small budget by substituting the cooperation and coordination of existing resources for capital. They are coordinated by the State Department of Agriculture, Division of Land Use, with support from a statewide federation of farmers' markets, and community sponsors. In some states, community organizations have taken the lead in establishing markets in their neighborhoods; in others, statewide groups or public agencies have initiated the markets.

If you are considering organizing a farmers' market, make certain that you give yourself enough time to organize properly. Planning for a farmers' market scheduled to open in July (a good target for first-year markets in the Northeast) should begin no later than January. Following is a timetable that can serve as a guide for your work.

January	1	Gather a group of interested people and call a meeting.
	2	Determine your specific goals and tasks.
February	3	Explore the nuts and bolts of each of these tasks.
		a. Locate a site for the market.
		b. Generate community support; raise funds.
		c. Check into legalities such as permits.
		d. Begin publicity to farmers.
March–April		e. Organize the management of the market.
May		f. Begin publicity to consumers.
July		g. Open the market.

Let's take a look at each of these steps outlined above.

Step 1: Gather a group of interested people and call a meeting. It is extremely important to involve other community groups in the planning process from the start. You should invite representatives from neighborhood associations, local chambers of commerce, church groups, and others to an initial organizational meeting. Use this meeting to generate interest in the idea of setting up a neighborhood farmers' market. Many people you invite may not be familiar with the concept, so be prepared to answer a lot of questions. It is important to outline the benefits of a market for your area, but be prepared to discuss any drawbacks as well so that problems are anticipated and solved, not left to undermine your efforts later. Any information you can get on other successful farmers' markets in the area will help give people a sense of what a farmers' market is and how it works.

Step 2: Determine specific goals and tasks. There are three aspects of a successful farmers' market to be coordinated: the farmers; the consumers; and the market site. You may want to set up a committee to work on problems related to each of these.

Step 3: Explore the nuts and bolts of each task. The first order of business should be securing a location for the market. Every market site is different, but there are some common factors in their success which should be kept in mind when looking at potential sites.

First, your market should be located in a place that is easy to find. High visibility is extremely important. You can have the greatest organization in the world and plenty of farmers, but if no one knows where the market is, it won't be successful. Second, the market should be accessible by public transportation if it is in an area where many residents rely on public transportation. If most residents have cars, be sure it has adequate parking close by. Third, the market should provide easy access for the farmers' trucks; and, fourth, it should be close to a business district. Lastly, it should be in an area zoned for retailing.

Make sure that the site has enough room to accommodate the farmers and their trucks, as well as shoppers and their cars. Find out who owns the site and contact the owner for permission to use the site for a farmers' market. Be sure to get this permission in writing. Also make the proper arrangements to ensure that the site is thoroughly cleaned up after each market.

After locating a site, the next step should be to look into the legalities of operating the market. Some states may require the farmers to have sales licenses or permits. Scales used to weigh produce may have to be tested and sealed, and there may be local health ordinances that apply to the market. Check with your local police and city or town hall. Inform them in writing that you plan to operate a farmers' market at the location you have selected, and ask if there are any formalities you need to go through in order to run the market. Check with the owners of the site to find out if they have liability insurance that would cover participants in the farmers' market. If not, you will have to provide such coverage.

At the same time that some people are looking into the legalities, others should be attempting to raise funds and build community support. Speaking before community groups can be a way to advertise the market as well as to recruit volunteers. The farmers' market can be operated without a lot of money; however, you will need some basic supplies such as materials for sign making, plastic garbage bags, and flyers for publicity. One alternative to fundraising is to approach local merchants and businesses asking them to donate some of the materials you need.

Publicity to both farmers and consumers is crucial. Use the resources in your community, such as local newsletters or newspapers, artists who can design flyers and a banner, and public or private meetings where announcements can be made. Many local newspapers will run a story on the farmers' market, particularly if they receive a good write-up from you. Local radio and television stations are required to air a certain number of public service announcements and could publicize the markets' schedule each week.

One of the most important tasks is to find someone to manage the farmers' market each week. The market manager should be involved in drafting market rules and is responsible for enforcing them during the markets' operation. A well-managed market, along with good publicity, will attract both growers and consumers and ensure a successful farmers' market.

Farmers' markets sponsored by community garden groups encourage the two types of growers to interact. A lot can be learned by talking with the professional growers whose livelihoods depend upon their agricultural practices. Conversely, farmers can learn more about their consumers. Many have been known to adjust their crops to respond to buyer preferences, rather than responding to the demands of wholesalers who may have other considerations. The community gardener and the small farmer often find, in fact, that they have many parallel concerns—about inflation and the lack of control over food prices, about the quality of food after it has been trucked across the country, and the lack of public concern for the unbalanced food production and distribution system. A farmers' market can mean that both farmers and community gardeners will begin looking beyond their own plots of land.

21

A Community
Canning Center

———◆·◆———

Susan Naimark

———◆·◆———

Canning is another natural extension of gardening—preserving the food that is grown. Once gardeners gain experience and confidence and begin to see the products of their gardens, they think about how to preserve those extra squash and tomatoes and how to put by the food they grow for later use. Garden produce can be dried, stored in controlled environments like a root cellar, frozen, or, using that traditional method of putting by, canned. But all of these methods, and particularly canning, require equipment, space, and skills that one gardener alone may not have. Canning lends itself to cooperative action, to gather, as a group, the resources needed to extend the garden beyond harvest right into the winter kitchen. Here is the story of how one group set up such a resource.

In August 1976, Massachusetts' first community canning center opened. Started by a county-based group called Women in Agriculture, Food Policy and Land Use Reform, Inc., the staff and equipment at the canning facility encourage gardeners and nongardeners alike to make the most of locally grown produce. The center provides the equipment for gardeners to put up their surplus, increasing their savings in food bills and enjoying homegrown produce year-round. Over eight thousand jars of produce have been preserved by over five hundred people at the center each year since its beginning. A canning center may be a logical next step for a community garden as its productivity increases over the years.

The Community Canning Center in Northampton, Massachusetts, operates out of donated space provided by Hampshire County. The center raised funds from private foundations to cover the cost of purchasing equipment and has used public funding to pay up to nine staff people. As public funds have become scarcer in recent years, however, the center has become more dependent on volunteer labor. Users are charged a small processing fee based on the number of jars of produce they process. This user fee covers all operating expenses beyond rent and wages.

Because the center operates on a self-help basis, the staff is prepared to provide as little or as extensive assistance as is needed. They have found that most users require considerable help. Clearly, the availability of the center has encouraged those who had preserved little or no food in the past to do so now. When resources allow, low-income users are assisted by a subsidized processing fee, provision of child care, and a produce-buying service that facilitates the use of food stamps in buying directly from local farmers.

Anybody can use the Community Canning Center simply by making an appointment in advance. Customers bring the ingredients and jars they plan to use, and the staff explains the use of the equipment to them. Jars are also sold at the canning center, which is open from May through December. The peak canning season is mid-August through mid-October, when the center is open six days a week, eight to fifteen hours a day. A farmer-to-consumer referral service provides a link between local growers and consumers who want to purchase produce in bulk. A related produce-buying service has been available when there has been enough staff to assist people who need transportation or who want to use food stamps.

A coordinating council of ten members, representing a wide range of community groups, develops policy and long-range goals and carries out fundraising efforts. Decisions affecting daily operation of the canning center are made by the staff as part of their commitment to worker control.

To operate the canning center and in order to expand into other projects, Community Self-Reliance, Inc., was formed in the first year of the center's operation. The goals of this organization are: (1) to support the local agricultural economy by stimulating a direct market outlet for local producers; (2) to provide local consumers with access to low-cost, locally grown, nutritious food and the means to preserve it safely for winter use; and (3) to educate the public about issues of food distribution practices and the possibilities for increased regional economic self-reliance. As Community Self-Reliance has grown, educational programs have been developed, including Project Greenbean, funded by the Federal Community Food and Nutrition Program, which in one year provided skills and information on gardening for food preservation to eighty low-income households, and From Seed to Table, a project funded by CETA, to research the existing local food distribution system and to develop a model for a locally controlled food system. Later, a public education program was developed based on the results of "From Seed to Table." A community resource center was created where materials are available on food, agriculture, nutrition, and community economic development. Work was begun to involve other groups and individuals to build a coalition for local self-reliance, including the formation of a community development corporation.

The Community Canning Center staff became interested, after its second year, in commercial canning to cover some of its operating costs. They found, however, that the volume they would need in order to be a successful commercial operation could not be processed with the equipment they were using for self-help, small-scale canning. They also found that extensive planning and organization would be necessary to establish sources of supplies, investigate legal and health questions, and determine markets for their products. Concern was expressed that commercial production would create a factory environment making staff jobs less desirable.

Since this initial investigation into commercial production, the canning center has moved into a larger space with the potential for expansion. The staff decided to experiment with some pilot volume production of various items during the months at the end of the season when the facility and staff were not as pressed for time. This pilot study included the marketing of each item for comparison of its success on the market. Commercial production is still in the experimental stage.

Some business planning has been done, and it is believed that such an operation could subsidize—although it would not completely cover—the operating costs of the canning center. Although still in its testing stage, the idea of a self-supporting canning center, controlled by its staff, could provide one more link in the network of local food self-reliance. Most regions of our country cannot grow all of the food needed regionally year-round. A canning center extends significantly a community's ability to provide for its food needs. It gives gardeners something to do with all of those tomatoes they can't even give away at the end of the summer. Canning greatly expands a garden's potential, allowing its produce to be enjoyed any time of the year.

22

A Food Co-op Community Garden

＊·＊

Susan Naimark

＊·＊

The Cambridge Food Co-op Garden, a garden created to serve an already established food co-op, grew out of the interest of several co-op members and the cooperation of two state agencies, the Fernald State School and the Massachusetts Department of Agriculture, Division of Land Use. The co-op garden became a reality in 1975, with the help of many hands and shovels. The garden was started by members who felt that a vegetable garden would not only supplement the food co-op's produce with organically grown local produce but would also serve as an opportunity to educate city people about how the food they eat is grown.

The co-op garden is located on the grounds of a state school for mentally handicapped adults, a twenty-five-minute drive from the food co-op. The land at the garden site was once farmed, and the produce was used at the institution. Now, approximately 2,500 square feet are being cultivated, producing well over a ton of produce each year for the food co-op.

The garden is a cooperative venture, and all of the planning and work is done by food co-op members as part of their required volunteer work hours. In February or March signs are posted in the co-op store to advertise the garden's start-up. A meeting is called for those interested in participating in the garden work. The volunteers have varying levels of knowledge about gardening, and many who have had little gardening experience use this opportunity to learn more about growing food.

A notebook containing a map with directions to the garden and garden work sign-up forms is kept at the store. Co-op members sign up in this book for a work shift at the garden to fulfill the monthly work hours required of co-op members. There are usually two work shifts per week, one on a weekday and one on the weekend.

These work shifts are led by a member of the co-op who is familiar with the garden and knows what tasks need to be done; usually several people have this familiarity and help coordinate the garden activities. At the garden, the coordinator for that day generally spends his or her time showing others what needs to be done and assigning new tasks as jobs are completed. Often the coordinator begins a work session with a tour for members of the work team who are new to the garden.

Many new gardeners find they very much enjoy the experience of touching the soil and seeing the plants grow. Co-op members who have been to the garden at least once are encouraged to return to fill more work hours at the garden. With a phone call to one of the coordinators, members can verify priorities and details of the work to be done and can then go to the garden at their convenience, recruiting others to join them if they choose.

Individual garden plots, in which members can grow their own food, adjoin the co-op garden. Because the garden site is some distance from the store, these personal garden plots have added to the incentive for members to make the trip. Having the co-op side by side with the individual plots also helps distribute the regular work of watering and weeding.

Two or three times a season there is a picnic/work day at the garden to which all co-op members are invited. A typical turnout is twenty-five to thirty people. These days provide new people with an opportunity to become involved in the co-op and are an enjoyable way of getting a lot of weeding, planting, and other work done. These days also give city children a chance to run under the sprinklers and play in natural surroundings.

In order to have any influence on the co-op's summer produce orders, the garden needs to provide a sizable quantity of each vegetable at one time. Therefore plants that yield heavily are preferred; peas, beans, tomatoes, and peppers have been the mainstay of the garden. Other easy-to-grow, high-yielding vegetables grown in the garden include lettuce, carrots, summer squash, and chard.

Among the gardening procedures used at the co-op are raised-bed planting, mulching with hay or composted leaves, green manuring (planting a crop that will later be turned back into the soil to add nutrients), and intensive planting of single crops. Many workers will garden in the co-op only once or twice, so intercropping is not practical; the layout of the garden must be clearly marked off with large sections planted in one crop. The co-op has built compost bins from wooden pallets; and, in addition to using manure and plant matter, the co-op relies on its own produce section to provide a constant source of organic refuse for composting. Individuals can add kitchen waste to the compost pile in exchange for small amounts of garden soil or finished compost for window boxes.

The garden is run without the use of chemical fertilizers or harmful pesticides. This is a preference expressed by many of the co-op's customers and is honored by the gardeners—even at the expense of forfeiting a portion of some crops to insects or of putting in additional time to pick insects by hand. Praying mantises and lady bugs have been introduced into the garden to prey on harmful insects, and the co-op practices other ecologically balanced gardening methods.

There is much value to having a community garden plot operated cooperatively. Gardeners who are not willing or able to make a long-term commitment to preparing, planting, maintaining, and harvesting a family-size garden take this opportunity to participate in an ongoing garden to the degree to which they are able. Crop rotation and succession planting are carried out more easily in a cooperative garden than in a garden made up of individual plots, as the entire area is planned each year as one unit.

The co-op garden's success has relied on a great amount of volunteer organization. The planting, garden maintenance, transportation, and harvesting must all be done according to a carefully planned schedule in order to succeed. The income from the produce cannot pay for the time required to grow it, but the value of fresh, organically grown produce goes beyond the market price that could be placed on it. In a cooperative organization where each member contributes work hours, a garden is an exciting way to utilize volunteer labor while contributing to the goal of increased local food self-sufficiency.

23

Permaculture
in Community Gardening

———◆•◆———

David Rosenmiller

———◆•◆———

The term *permaculture,*[1] developed by Tasmanian agriculturalists David Holmgren and Bill Mollison to define an entire agricultural system, suggests a combination of permanence and agriculture: the creation of a permanently sustainable
agricultural system.[2] Permaculture is a system using tree crops, perennial plants,
animals, and buildings, in addition to annual vegetable crops, to form integrated
functional design. High yields are produced with a minimal expenditure of energy
(including human energy), by weaving together the above components and working
with the sun, wind, and rain. For example, permaculture uses perennial plants
that serve several purposes at once, such as improving the soil, keeping down
weeds, deterring pests, and providing food.

The idea behind a carefully planned, integrated agricultural system is that the
benefits are greater than if the placement and choice of plants were done randomly
or segregated as they are in a monoculture (single crop) system. Interdependent

webs are created intentionally among the plants, insects, animals, and soil, duplicating the relationships found in nature, where productivity is high. Such designs create ecologically stable, produce-yielding landscapes that are as beautiful as they are productive.

Permaculture is applicable to many different types of agriculture, from large- to small-scale gardens. Because it can improve an area ecologically, permaculture is especially useful in disturbed ecosystems such as vacant lots.[3] The use of permaculture in community gardens provides us with a strong vision for the future. It may include the use of fruit and nut trees, fences covered with roses and berry bushes, beds of perennial vegetables and flowers, evergreen hedges, windbreaks, shaded sitting areas under arbors of grapes, and flowering vines. These are just some of the possibilities, and all of them have more uses than meet the eye.

THE ELEMENTS OF A PERMACULTURE SYSTEM

Perennial Plants

Perennial, or self-perpetuating, plants, are central to permaculture because they perform many important functions. Several types of perennial crops—fruits, nuts, vegetables, herbs—are grown for food. Other plants are grown primarily for their usefulness in developing the ecological web of a garden but have edible leaves, roots, or stems as well. Perennial plants provide windbreaks, attract beneficial insects that prey on pests, attract pollinating insects, lure or decoy pests away from the food crops, improve the soil structure and its nutrient content, retain topsoil and water, reduce air, water, and noise pollution, deter vandals, recycle wastes, and control weeds. Perennials are especially important because they don't require tilling, which exposes the ground to erosion, water loss, and weeds each year.

Microclimate Modification

Optimal growth can be achieved if you choose plants that grow well in your location and soil conditions. You can also create beneficial microclimates (the unique weather conditions of small areas) for the plants you want to grow. This can be done by planting where walls, buildings, trees, or hedges catch the sun's rays and deflect the wind. Such strategic planting allows plants to thrive that under less sheltered conditions might not do so well.

Multifunctional Plants

One element of permaculture design particularly important to community gardeners is the use of plants with more than one function. A climbing rose grown on a garden fence can deter vandals, provide nutritious rose hips and petals, make the garden more attractive, attract beneficial insects (for pollination and pest control), and shade plants that do best in moderate sun. Because properly chosen plants serve several purposes, they enhance the community garden without increasing maintenance needs.

Soil Improvement

Good soil is essential to all plant growth and, consequently, to all life. Exposed soil does not appear in a balanced natural environment; it is damaged soil and susceptible to further damage by the elements. Weeds, however unlikely it may seem, are nature's way of improving damaged soil and restoring ecological stability. Weeds protect, aerate, and add nutrients to the soil. Weeds are, by definition, simply unwanted plants, but they can be useful under the proper circumstances. Weeds can help loosen the compacted soil often found in vacant lots and new community gardens.

All organisms have many functions in their natural settings. Understanding some of the functions of "weeds" enables us to use these plants to our advantage. The key to permaculture is the careful observation of how plants and animals interact, creating an integrated, mutually beneficial environment.

Rather than cultivating and turning the soil over—which damages and exposes it to the elements—permaculture uses alternative cultivation practices that aerate the soil and lift it only slightly. In most types of agriculture, a plow or rototiller is used to turn over compacted soil. Using permaculture techniques to aerate the soil, a heavy pitch fork is used to gently loosen the soil without turning it over. Herbaceous plants with long tap roots such as comfrey, Daikon radishes, alfalfa, and chicory can be planted in the spring and mowed before they go to seed. Let the plant and its roots decay to provide air and water channels in the soil.

Trees and shrubs are a more permanent means of conserving and improving the soil—once adequate soil exists in which they can grow. Varieties that tolerate poor soil conditions may need to be given priority.

Cover crops (living, green mulches) such as alfalfa, clover, buckwheat, and vetch can be planted in the fall and early spring where there are no other crops (pathways, around the base of trees) or for a year prior to other planting. These cover crops are legumes, which provide the soil with nitrogen by means of the nitrogen-fixing bacteria that grow around their roots. Legumes are especially valuable when planted among young fruit trees, as fruit trees require nitrogen during their early development. The blossoms of some of these plants play an important role in orchard pest control and may provide nectar for bees to live on during the long period when fruit trees are not in blossom. Everything in permaculture design has several functions.

Sheet mulching is another practice recommended by permaculturalists for improving the soil in small vegetable plots. Simplified, this is a thick layer of mulch, such as newspaper, laid down over the ground, weeds, and grass. On top of this is placed more mulch material such as manure, leaves, seaweed, sawdust, or straw. Vegetables are then grown in clumps of soil placed on top of this sheet mulch, or in holes cut through it. It is a system that requires little water and labor, suppresses weeds and creates excellent soil. The mulch layers slowly break down, adding nutrients to the soil below, much like compost. This technique could be alternated yearly with direct planting, or to start a new bed.

Mulching in general is an important practice in permaculture. Leaves, wastes from perennial and annual plants, and other organic products (manure, sawdust, wood chips) can all be used. The addition of well-rotted compost and earthworms will also improve the soil. Typically, however, most compost would be reserved for greenhouse or container growing.

Nutrient Cycling

Soil in vacant lots is often of poor quality. Using plants, it is possible to improve the soil structure and nutrient content with less emphasis on traditional methods of cultivation and fertilization. This process is called nutrient cycling. Certain plants that accumulate nutrients otherwise inaccessible to most crops—from the atmosphere and from the soil and the subsoil—make them available to plants they grow near.

> Most legumes provide essential elements in this way, as may any vigorous and deep-rooted tree or weed which grows below the upper, already leached layer of soil. All material in nature cycles via wind, water, dust, and human or other animal activity. Some plants and some animals act as catch-nets for rare and essential elements and can be used in any garden for that value alone.[4]

Many herbaceous (plants without woody stems) perennial plants accumulate nutrients as they grow. These include borage, comfrey, lambs quarters, purslane, plantain, chamomile, dandelion, dock, stinging nettle, vetch, watercress, and yarrow. When these die back each year their leaves and stems decompose, supplying the soil with nutrients and organic matter. Gathering and composting them accelerates this process.

Tiered Planting

In permaculture design beneficial plants are placed under and near tree crops to help provide them with nutrients. The trees in turn modify the microclimate for these herbaceous plants, providing them with protection from the sun, wind, and rain. This mutually helpful relationship is an example of tiered planting, where space and sunlight are used more efficiently by plants of different heights growing in the same area.

Many trees, including fruit trees, should not be planted directly adjacent to vegetable crops because they will compete for the same nutrients. An intermediate zone of useful and edible perennial plants can act as a buffer between trees and annual vegetable crops, while improving the productivity of each. The north side of a garden can be bordered by a row of trees, with plants among and in front of the trees and vegetable plots to the south. Similarly, a pathway with a green mulch (of clover or alfalfa, for example) will also serve this purpose. Later the cover crop can be cut and used in the garden as mulch.

Permaculture planting in a community garden.

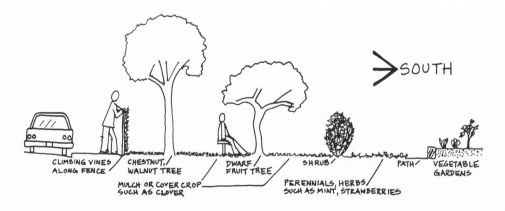

Tree Crops

Trees are important everywhere but nowhere more than in the city. Trees purify the air, reduce air and noise pollution, humanize built-up streets, and symbolize life and growth. Trees with edible produce are doubly important because of their function of providing food, increasing local food self-sufficiency.

Many kinds of fruit and nut trees as well as berry bushes are suitable to each region of the United States. Urban community gardeners have a climatic advantage because cities are generally warmer than outlying areas. By protecting tender trees from the elements with walls, buildings, or other trees, varieties can be grown that otherwise wouldn't survive. The best way to determine which tree varieties are suitable to your area is to contact your local county extension service, local nurseries, or other professional growers.

Dwarf trees, which reach heights of six to eight feet, are available in most varieties of fruit tree and offer gardeners a number of advantages. Four dwarf trees take up the same space as one standard sized tree. Dwarf trees are easier to protect against pests, easier to prune, and easier to harvest because a ladder is not required. The trees begin to produce fruit earlier than their normal-sized relatives, usually in two to three years. If vandalism is a problem, dwarf fruit trees offer an alternative to broken limbs as they are too small to climb. However, dwarf trees should be protected from careless damage such as damage from vehicles, as should any tree. Dwarf trees may not be ideal for every urban situation, and each site should be evaluated according to its unique characteristics.

Perennial Fruits and Vegetables

An important step in establishing a permanent, diversified garden is the inclusion of perennial vegetables, herbs, and fruit. Garlic, perennial onions, rhubarb, asparagus, Jerusalem artichokes, mint, rosemary, sage, thyme, chives, and marjoram are all perennials in certain parts of the country. Perennial vegetables may be grown in beds separate from those growing annual vegetables if you rototill or turn the soil in your vegetable plot every year.

Many fruit shrubs also will grow well in community gardens. Strawberries and low-bush blueberries, high-bush cranberries, elderberries, currants, grapes, blackberries, raspberries, and bush cherries will all work well as hedges. Some fruit shrubs have the additional advantage of producing beautiful flowers or sharp thorns to protect against vandalism.

A hedgerow can shelter tender plants, both annual and perennial, grown beneath or next to it. Dense hedgerows can be grown from small trees or tall shrubs. Hawthornes and lucernes are examples of trees that, with proper pruning, can eventually grow to form a contained yet impenetrable barrier.

Biological Pest Management

Pest management in a permaculture system is a matter of protecting plants on several levels. Healthy soil and healthy plants are generally much more resistant to pests, diseases, and fungii. The use of resistant varieties enhances the ability of trees and plants to protect themselves when well cared for.

In nature, pests are controlled by predatory insects and birds. Without these natural enemies, insect populations would be uncontrollable. In permaculture this effect is enhanced by attracting the predators of harmful insects to the garden with the perennial plants they prefer. These are called insectory plants and include tree of heaven, coyote brush, rue, fennel, wild buckwheat, euonymous japonica, salt-brush, soapbark tree, and California coffeeberry. Each region has many indigenous plants that attract birds; examples of these include blackberries, crabapples, and rugosa roses. Some plants that attract beneficial insects such as lady bugs and small, nonstinging wasps are borage, comfrey, lucerne, fennel, mint, and clover. Lady bugs and other predatory insects can also be purchased through mail order catalogs,* and introduced into the garden.

Some perennial plants repel pests naturally by mechanical or chemical means. Others act as decoys, luring pests away from food crops. Trees with berries that attract pest-eating birds may also lure birds away from your food crops. A bird-house provided near such trees will encourage the birds to stay.

Plant diversity is an important element of pest management. Various species of trees and crops are interplanted, forming a polyculture (as opposed to mono-, or single crop, culture). Some trees will act as hosts for the predators of harmful insects, at the same time providing food. Planting a variety of crops lessens the chances of the destruction of a single crop from an outbreak of a particular pest. Because of their checkerboard design and the diversity of their crops, community gardens benefit from this effect already.

It takes considerable observation of the pests in one garden to design planting that will protect crops against those particular pests. Integrated pest management programs are increasing in orchards throughout the United States. On a small scale, when commercial return is not a factor, these programs should be even more successful.

*IPM Practitioner, Volume III, No. 5, May 1981, Route 1, Box 28A, Winters, California, 95694.

Permaculture Design in a Community Garden

Before applying the principles of permaculture to a community garden, it is important to get an overview of the site. First, identify the climate and microclimates of the garden. Look at factors such as the average annual temperature, rainfall, prevailing winds, available sunlight, and patterns of shading. Then make an analysis of how the land is currently being used, noting planting patterns (both natural and cultivated), traffic patterns in and around the garden, the condition of the soil and subsoil, and local vegetation.

Next look at the needs of the gardeners and the problems the site has, and think about what kinds of plants can solve these problems. This will require much close observation. If there are certain persistent pest problems, ask yourself, Why? If there is erosion, again, Why? Of the plants that can solve these problems, think about which ones will grow at your site and what beneficial functions each performs. How can microclimates in the garden be modified to better provide the right conditions for these plants?

Make a list of the qualities most needed in plants and the conditions in which they must survive. Then list the plants that will grow under these conditions and the beneficial qualities of each. This will require some research at your local library, horticultural society, or landscape nursery. The information you find will help you put together a site plan for the garden that meets the needs of your gardeners.

You may want to introduce permaculture into your garden by adding only one or two new elements to your garden each year. In this way you will be able to see how each element fits into the workings of the garden without disturbing the established arrangement. The evolving process of creating a community garden based on the theory of permaculture may parallel the evolution of the gardening group. The physical and organizational needs of a community garden both require careful analysis and attention to the individual components that make up the whole. And both should be striving toward the same goal: a permanently sustainable agricultural system.

24

Harvesting
Our Experiences

◆-◆-◆

Pat Libby

◆-◆-◆

This business of urban gardening in some areas of the city is treated as an artsy-craftsy kind of isn't-it-nice-gravy-on-the-potatoes latter day outgrowth of the touchy-feely sixties, as opposed to the way in which it's looked upon in my neighborhood certainly, and in places in East Boston and a number of other parts around the city, where the participants look at it as a focus, as an arena, as a forum for working with their hands, producing food, something that has inherent value, something that sets one outside of the definitions for work by the larger society, as a place where cross-generational contact naturally begins to happen, as a place where you begin to see rustlings and arguments and struggling going on around appropriate agricultural technology.

—Ray Almeida, community gardener

One sunny June day in 1976, a caravan of twelve National Guard trucks wound their way through Boston's South End and dumped loads of donated topsoil on two vacant lots. The very next day, two more lots were filled with miniature mountains of trucked-in topsoil, and members of the South End Garden Project rejoiced at their first victory. One of the garden organizers later wrote that she "had agreed to be responsible for bringing the topsoil into the city and felt as if she had just promised to learn to fly." With no funding whatsoever, the vision shared by an urban neighborhood—of empty lots sprouting community vegetable gardens—became reality.

Within a six-month period, neighborhood volunteers had solicited donations of topsoil, the use of seven additional city lots, transportation, heavy equipment, seeds, plantings, landscaping materials, fertilizers, and water—worth in aggregate well over $12,000. More than 100 community residents became involved in planting gardens that first year, and in 1977, the number of gardeners in the South End doubled. Across the city, similar efforts were underway. Blighted lots strewn with garbage and filled with weeds were transformed into neat rows of colorful vegetables. Garden organizers around the city began meeting to discuss a unified strategy for solving common problems and shared goals. By the turn of the year, Boston Urban Gardeners was born.

The BUG coalition, incorporated in 1977, consisted at first of members of three Boston communities: the South End, Roxbury, and Jamaica Plain. Through hard work and persistence, the group secured funding for two part-time staff members, initiated urban soil toxicity research, began negotiating with the city concerning site permanency, scrounged and distributed resource materials, sponsored workshops, worked with agencies and community groups to support new garden development, published a newsletter, and worked tirelessly to set in place the cornerstones of the urban agricultural network that now exists in Boston.

The original components of BUG's citywide program remain the same today as in 1977; they include

1. *Education and information.* Facilitating communication among gardeners through a newsletter, bimonthly coalition meetings, and periodic workshops and conferences.

2. *Garden organization and site development.* Facilitating the process of locating land, attracting potential gardeners, grading, fencing, bringing in topsoil, supplying water, laying out plots, and assisting in locating donated or low-cost materials and supplies.

3. *Technical assistance.* Providing on-site instruction and help with planting, cultivation, pest control, and construction of such garden aids as cold frames and compost bins.

4. *Research and development.* Answering questions about toxic pollutants, growing techniques, alternative technologies, and soil improvement.

5. *Composting.* Converting selected urban organic wastes, such as leaves, supermarket trimmings, and horse stable manure, into compost (or humus) for gardens in order to reduce dependency on expensive commercial sources in distant suburbs.

6. *Fundraising.* Helping members to obtain the funds necessary to implement their work in urban gardening.

Planted in 1976, the BUG network has stretched its roots and grown longer branches. Boston now enjoys more than 120 community gardens ranging from small informal garden groups to incorporated, tax-exempt organizations with paid staff and membership numbering in the hundreds. The goals shared by all of these groups are to strengthen the local food system, reduce neighborhood blight, and create an environment that supports cooperation and respect across ethnic, racial, and economic boundaries.

MOVING TOWARD A GREENER TOMORROW: INSIGHTS ON INDEPENDENCE

Since BUG's establishment, we have observed that, within a few years of starting a community garden, the people involved begin to think about projects that tie their garden group into a wider network. Gardeners work collectively to accomplish what they are not able to do individually. Small businesses are developed to support the gardens. Community control of other neighborhood open space is discussed and pursued. As garden groups evolve into organizations concerned with the overall quality of neighborhood life, gardeners begin to perceive the enormity of their impact on the community.

"Mighty oaks from little acorns grow," was how Ed Cooper, BUG's president since 1978, recently described the growth of the organization. "BUG has been a terrific catalyst in the development of urban gardens within the City of Boston. There are still thousands of lots scattered all over Boston which are growing up into weeds which could be developed into community gardens."

Ed Cooper's garden is but one example of what can happen to such a lot. Five years before his retirement in 1974, Ed decided that his postemployment goal would be to work with senior citizens. At age seventy, he surveyed his neighborhood of Highland Park in Roxbury, counted 400 senior citizens living in the area and formed the Highland Park 400.

Around the corner from Ed's house stood a vacant lot measuring 8,000 square feet and featuring two burnt-out house frames. Ed and four other senior citizens dug up the tillable soil and planted four gardens. The following year, he called the City of Boston and convinced them to remove the houses and to fence and cover the area with eight inches of topsoil. Now, there are twenty-three members of the Highland Park 400 gardening on fifteen-by-thirty-foot plots on the site. Moreover, arrangements are now in the making for securing land adjacent to the site where there will be a farmers' market, rose garden, seating area, fruit and nut orchard, canning facility and greenhouse.

This garden exemplifies BUG's belief in community development through neighborhood efforts. What began as one man's vision of an urban garden has evolved into plans reaching far beyond the original concept.

The struggles and dreams of community gardeners all point in the same direction: self-sufficiency. One of Boston Urban Gardeners' fifty board members, Julia Brown, who organized several community gardens in her neighborhood, has not purchased a single vegetable in three years. She has been able to grow what she needs for her family of five and has purchased a freezer to preserve her fresh produce for year-round use. Booker DeVaughn, another BUG board member, has been fighting to preserve his community garden—nine acres of cultivated land gardened by over 435 Boston families—for permanent agricultural use. This large garden is located on state property, and legislation has been drafted to maintain the site for community gardening.

Another group in Boston, Dorchester Gardenlands Preserve, started its first community garden in 1977. A year later the founding members of DGP organized a farmers' market in a nearby shopping area as another way of providing fresh food to city residents. The success of the farmers' market stimulated the group's interest in direct marketing of produce. At a Washington, D.C., conference on direct marketing, a DGP member met a California orange grower. In the winter of 1981, faced with an unmarketable surplus, the grower donated three 40,000-pound shipments of oranges to the group. DGP paid the packaging and shipment costs and developed a local distribution system to hospitals, universities, school systems, and social service organizations. Today, the distribution system is expanding to include produce purchased at Boston's wholesale market. This project employs three people and in the future will seek to market produce grown by Massachusetts farmers. What started as one small community garden has expanded into a community development corporation with three garden sites, a weekly farmers' market, and a food distribution business.

The Southwest Corridor Community Farm, another organization in the BUG coalition, began as a publicly funded (CETA) employment project. In its first year, a one-acre community garden was established, and a 600-square-foot solar greenhouse constructed. For four years the greenhouse has produced spring seedlings for sale to urban gardeners. Initially staffed by volunteers, the seedling production now pays two gardeners to maintain the operation. The BUG network has been utilized to advertise the seedlings, and gardeners from throughout the city have come to depend upon the farm each spring. Recently, the farm has acquired a second garden site and is expanding its greenhouse seedling business.

BUG, too, is initiating its own business venture: a composting business to turn local organic waste into topsoil. During 1978 staff from BUG, the Massachusetts

Department of Food and Agriculture's Division of Land Use, and a team of student interns conducted a feasibility study of a large-scale composting facility. Preliminary study findings encouraged the group to secure funding for a physical demonstration project, which took place during the summer of 1979. The results have been published in a report to the project's sponsor, the New England Regional Commission, entitled "Large-Scale Urban Composting of Vegetable Wastes: A Pilot Project, 1979." Subsequently a business plan for the operation was drawn up, and efforts to raise start-up funds are currently underway. The compost facility will supply a steady source of high-grade, affordable topsoil for local gardeners. Recycling local resources while providing a much needed alternative method for the disposal of solid waste in Boston, the facility will generate and recycle revenue within the community and create employment opportunities as well.

These experiences demonstrate the potential of community gardening as a powerful, effective means of stimulating community economic development. The BUG coalition's goals led to increased community self-reliance by providing local jobs as well as local competence in the provision of basic needs. Other coalition projects include land acquisition, nutrition and gardening educational programs, audiovisual presentations, legislative advocacy on food and farm issues, and the development of model landscaping designs for permanent community gardens.

This book is a glimmer of our collective knowledge of community gardening. We hope it has offered inspiration and encouragement. Sharing in the creation of living open space engenders a spirit of community—a common unity—that cannot be fully described in words. Real learning is accomplished by doing and discovering; in this case, by getting your hands dirty.

The benefits of affordable nutritious food, wholesome recreation, and a feeling of self-reliance and competence are universal consequences of planting a successful community garden. At a time when inflation is rampant and vacant lands abound in cities and suburbs, producing our own food allows us to exercise a positive influence on our surrounding environment and on our lives. BUG has found—and hopes readers of this handbook will agree—that the achievement of organizing and planting a community garden far exceeds the value of the produce harvested.

Tables

————— ◆•◆ —————

Table 1
COST ESTIMATES FOR A COMMUNITY GARDEN

One-Time Costs	Low Estimate	High Estimate
Incorporation	$ 30	$ 200 (includes legal fees)
Land acquisition*	100	5,000
Topsoil, delivered (at $8.50–13 per cubic yard)	400	2,000
Soil amendments (limestone, granite dust, peat moss, greensand, etc.)*	50	250
Water installation		
Meter	250	500
Copper supply line	800	2,000
PVC (plastic) line	50	250
Fencing, installed		
6′ chain link ($6.50 per linear foot)	2,000	6,500
6′ wooden ($4.00 per linear foot)*	1,200	4,000
turkey wire (do-it-yourself)	300	2,000
Landscaping		
Planting materials*	25	500
Railroad ties ($2–12 each)*	40	1,000
Wooden stakes*	40	100
Building projects		
Compost bins*	10	75
Benches, tables*	25	1,000
Tool shed*	100	1,000
Special activities		
Entertainment*	25	500
Educational*	10	500
Total	$1,455	$20,955

*Available free, recycled, or donated.

163

Table 1

COST ESTIMATES FOR A COMMUNITY GARDEN (*Continued*)

Ongoing Costs (on a yearly basis)		
Taxes	$ 75	$ 1,300
Water bills ($2–4 per plot)	50	200
Garden coordinator*	100	600
Bookkeeper*	100	1,500
Mailings		
Photocopying*	5	100
Postage*	15	150
Miscellaneous office supplies*	20	150
Administrative (typing, phone calls, etc.)*	10	200
Travel (gas, truck rental)*	10	200
Liability insurance	55	100
Manure*	5	50
Tools*	30	100
Maintenance		
Fence repairs, chain link*	50	1,000
Fence repairs, wooden*	30	200
Water line repairs, PVC (plastic)*	5	100
Water line, spigots or standpipes*	50	1,000
	$ 510	$ 6,650

*Available free, recycled, or donated.

Table 2

RECYCLING RESOURCES

Item	Possible Uses	Likely Sources
Vegetable waste	compost	farmers' markets, home and restaurant garbage
Plastic gallon jugs	passive irrigation, scoops, watering cans (punch holes in lid)	trash
Fish crates	mini-compost bins, raised beds, scrap lumber, seating	local fish markets
Bushel baskets	containers for carrying produce, planters	local markets
Horse manure	composting/fertilizer	stables
Plastic 55-gallon drums	watering systems	chemical supply houses, natural food stores, food-processing plants
Old, leaky hoses	watering systems	trash, last year's garden
Carpet scraps	mulching/paths, weatherstripping for cold frames	trash, carpet companies
Old cribs	(whole crib, legs and springs removed) turn upside down and use for composting (sides raise for removal) (springs and/or sides) trellis for climbing crops	trash, junkyard
Wooden pallets (skids)	compost bins, fencing	companies receiving deliveries on pallets
Old tires	self-heating raised beds, (the black absorbs heat from the sun), planters (cut from rim on one side and turn inside out)	garages, trash, junkyard

Table 2

RECYCLING RESOURCES (*Continued*)

Item	Possible Uses	Likely Sources
Mop and broom handles	staking	trash, junkyard
5 gallon buckets	carrying water, mulch, produce, etc.	trash, restaurants, painting contractors
Cracked plastic trash cans	compost bins	trash, junkyards, hardware stores
Black and white newspapers (colored ink contains lead)	mulch	trash, newspaper vendors
Scrap lumber	compost bins, raised beds, cold frames	wrecking crews, building contractors, hardware retailers or wholesalers
Used plastic dropcloths	mulch, cold frame, greenhouse	painting contractor
Fallen trees	raised beds, terracing, children's climbing equipment	park and forestry service, city tree service
Large packing crates	compost bins, play equipment for children, storage for tools, tables	plumbing/electrical supply, house appliance retailers
Milk crates	raised beds, planters, seating	markets, school cafeterias, milk distributors
Styrofoam cups	seedling pots, cutworm collars	trash, offices
Milk cartons, paper cups	seedling pots, cutworm collars	trash
Egg cartons	seedling pots	trash, supermarkets
Aluminum pie plates	scarecrows	trash, bakeries
Old window screens	food dryer panels	trash
Old storm windows, old frame windows	greenhouse and cold frame glazing	trash, building contractors
Old wooden doors	cold frame sides, compost bins	trash, building contractors, junkyards

Table 2   167

RECYCLING RESOURCES (*Continued*)

Item	Possible Uses	Likely Sources
Old fluorescent light fixtures	indoor light gardens	trash, building contractors
Old tin cans	scarecrows (remove labels)	trash
Plastic gallon jars	subterranean irrigation system	restaurants, school cafeterias
Glass gallon jars	mini-hothouse (pierce lid)	restaurants, school cafeterias
Old ribbons	scarecrow flyers	trash, sewing or fabric stores
Dried leaves	mulch, compost	local parks department
Bricks, cobblestones	borders for gardens, paths	old building sites, brickyards

Table 3

SOURCES OF FUNDING FOR COMMUNITY GARDENS

Source of Funding	Estimated Income	
	Minimum	Maximum
Individual donations of materials, labor	in-kind	unlimited potential
Donations of money from neighborhood businesses, business associations, local charitable and service organizations	$ 10	$ 100 each
Membership fees ($1–5/year per person or family)	50	1,000
Bake sales, fairs, raffles, other community events	50	1,000 per event
Public relations departments of corporations and banks	50	1,000
Local private charitable foundations	500	15,000
Regional or national charitable foundations	1,000	50,000+
Government program grants or contracts	2,000	50,000+

Table 4
GUIDELINES FOR PROPOSAL WRITING

I.   Summary
     One page or less: summarize project, the need for it and its importance;
     highlight its unique points, if any.

II. Background
     Describe your organization, its history, membership, constituency, and
     goals; briefly describe its accomplishments, especially any that are unique
     to your organization; describe its relationship with other community
     groups it has worked with or from whom it has received support in the
     past.

III. Assessment of Needs
     Define the problem(s) you want to solve with this project. Be specific—
     include statistics; state exactly what aspect of the problem you are taking
     on, be sure this relates to the description of your organization (in part
     II), and show clearly why and how your group is qualified to tackle the
     problem you are defining.

IV. Program Objectives
     Describe what you want to accomplish in specific, measurable terms. For
     example, the objectives of this project are: (1) to establish two new
     community gardens on land that is presently vacant in Lower Roxbury;
     (2) to start 75 low-income families gardening on the new sites; and (3) to
     provide general vegetable gardening instruction to 150 Lower Roxbury
     residents during the summer of 1983.

V.   Methods
     Step-by-step, describe activities you will carry out to accomplish your
     objectives. This can be in the form of a month-by-month time line or a
     chart that lists problems in the lefthand column, objectives after each
     problem in the middle column, and methods after each objective in the
     righthand column. Include an explanation of why you are using these
     particular methods (background research, past experience of your or
     another group); make it clear that you have thought through various
     alternatives. Identify groups you plan to work with or community
     resources you plan to utilize; mention any that have offered to donate
     materials or labor.
     (If your proposal is a small request, for funds rather than for staff or for
     subsidy of more elaborate projects, you can stop here and go on to Part
     VIII, Budget.)

Table 4   169

GUIDELINES FOR PROPOSAL WRITING (*Continued*)

VI.   Evaluation

A good evaluation process should accurately determine the effectiveness of your project in meeting its objectives; it should provide for feedback, as you go along, which will be necessary in determining how to adjust objectives, timetables, or material and staffing needs. Describe the evaluation process you will use, including your criteria for evaluation, which should be based on your objectives (Part IV Program Objectives). The evaluation process may include any combination of the following: weekly or monthly staff meetings, questionnaires administered at the beginning or end of an activity, regularly scheduled discussions with project participants. An ongoing evaluation process should provide the information you need for writing a final report at the end of your project as well.

VII. Other Funding

This section is *not* necessary if you are asking for a one-time grant, such as for initial garden development or for purchasing equipment. It *is* necessary if you are requesting program money for activities that will continue beyond the initial funding period. Describe how you plan to continue the project when the initial funding runs out; this may include a list of other places you plan to apply for funding or plans for generating funds through the project itself (charging for services, selling seedlings, holding annual community fundraising events, etc.). Funding sources want to see that a project they sponsor does not just fold after their funds run out.

VIII. Budget

This is the most important—and most closely scrutinized—part of any proposal; be sure that all costs have been thought through and are realistic. On looking over the following lists of possible costs, disregard any costs that do not apply to your project. You may not need insurance, for example, or office space rental for a garden beautification project. *Administrative Costs.* These may be absorbed by an existing organization, or administration work be volunteer. These costs cover items that don't go *directly* into the project but may be necessary as support, and include:

  1. Wages and salaries of support staff (bookkeeper, typist, administrator)

2. Fringe benefits
3. Office space rental
4. Office supplies
5. Telephone
6. Travel (mileage, vehicle rental)
7. Insurance
8. Postage
9. Photocopying

Project Costs. These will include some or all of the following:

1. Consultants and contract services (garden coordinator, fence installer, etc.)
2. Equipment rental or purchase
3. Project materials (include anything permanent, such as fencing, water systems, and compost bins)
4. Consumable supplies (seeds, stakes, paper, postage)
5. Travel (directly related to project activities, such as dumptruck rental to pick up manure)

You may be asking for funding to cover project costs only, especially if you are a new organization without paid staff or if you are working with an agency that can provide administrative support. Often administrative costs are included as a percentage (10–15 percent) of the project budget rather than itemized.

Table 5

COMMON NATURAL FERTILIZERS

Material	Nutrient Analysis			Immediate Availability	Comments
	Nitrogen	Phosphorus	Potassium		
Dried blood	12.0	0	0	high	may contain lead
Bone meal	2.0	20.00	0.20	moderate	may attract pests
Rock phosphate	0.0	20.00	0	low	must be ground to 200 mesh powder
Fish emulsion	4.0	1.00	1.00	moderate	may attract pests
Fish meal	10.0	4.00	0	moderate	may attract pests
Leaf mold	1.0	1.00	1.00	moderate	probably the most readily available material at the best price (free)
Seaweed	1.5	0.70	5.00	moderate	may be used fresh or dried
Cottonseed meal	7.0	2.50	2.00	high	may contain pesticide residue
Wood ashes	0	2.00	5.00	high	hardwood ash preferred
Garden compost, not dehydrated	0.5–1.0	0.25–0.50	0.5–1.00	moderate	quality depends on ingredients
Rotted manure, not dehydrated					should not be put in the garden when fresh
Cow	0.5	0.25	0.50	moderate	
Horse	0.7	0.50	0.60	moderate	
Rabbit	4.0	3.00	1.00	moderate	
Hen	1.8	1.00	0.50	moderate	
Hog	0.3	0.30	0.45	moderate	
Sheep	1.0	0.35	0.50	moderate	
Sludge	4.0	2.50	1.00	moderate	contains toxic metals
Granite dust	0	1.00	6.00		virtually insoluble
Lime	—	—	—		used to raise pH
Dolomite lime	—	—	—		corrects magnesium deficiency
Peat moss	—	—	—		used for organic matter content

Source: Suffolk County Cooperative Extension Service, University of Massachusetts, "Natural Fertilizers in the Home Vegetable Garden."
Note: Dash indicates negligible level

Table 6

THE FIFTEEN MOST NUTRITIOUS VEGETABLES YOU CAN GROW

Vegetable	Excellent source of vitamin	Good source of vitamin	Minerals
Green beans	A	B, C	calcium
Broccoli	A, C	Bs	calcium, phosphorous, iron
Beet greens	A	B, C	calcium, iron, potassium
Collards	A	C, Bs	calcium
Dandelion greens	A	C, Bs	calcium, sodium, iron
Fava beans	protein	Bs	phosphorous, iron (may cause allergies)
Garden cress	A, C	Bs	iron, potassium
Lettuce	A, C		calcium
Lima beans	protein	Bs	iron
Mustard greens	A, C	Bs	calcium, iron
Green peas		Bs, C	iron, phosphorous
Chili peppers	A, C	Bs	iron, potassium
Soy beans	protein	Bs	iron
Spinach	A	C, Bs	iron, sodium
Winter squash		A	riboflavin

Notes

———◆◆———

Chapter 1

[1]The Gallup Organization, Inc., 1979 National Gardening Survey (Burlington, Vt: Gardens for All, 1979).

[2]U.S. Department of Commerce, *Consumer Price Index: Fact Sheet* (April 1980).

Chapter 2

[1]James Drake. *A Picture of Birmingham: 1825*. Excerpted by H. Thorpe, E. B. Galloway, and L. M. Evans in *From Allotments to Leisure Gardens*. Birmingham, England, 1976, p. 2. Quoted by Mary Lee Coe, *Growing with Community Gardening*. Taftsville, Vermont: The Countryman Press, 1978, p. 11.

[2]Mary Lee Coe, *Growing with Community Gardening*. Taftsville, Vermont: The Countryman Press, 1978, p. 12.

[3]Thomas J. Bassett, "Community Gardening in America." *Community Gardening, A Handbook*. New York: Brooklyn Botanic Garden Record, vol. 35, no. 1, spring 1976, p. 4.

[4]Charles Lothrop Pack, *The War Garden Victorious: Its War Time Need and Its Economic Value in Peace*. Philadelphia: J. B. Lippincott Co., 1919, pp., 1, 10.

[5]Ibid., pp., 16, 17.

[6]Thomas J. Bassett, *op. cit.*

[7]Ibid., p. 6.

[8]C. Keith Wilbur, *The New England Indians*. Chester, Conn.: The Globe Pequot Press, 1978, p. 30.

Chapter 4

[1]Newberry, Rebecca S. *How to Organize a Community Garden in Massachusetts*. Boston: Massachusetts Dept. of Agriculture. Publication #11,327-46-350-5-79-CR. 1979.

²*Random House Dictionary of the English Language* (New York: Random House, 1966), p. 321.

³Parker, Allen L. "Interactive Networks for Innovational Champions: A Mechanism for Decentralized Educational Change." Thesis published in 1979, Harvard Graduate School of Education, p. 38.

⁴*Random House Dictionary of the English Language,* p. 1284.

Chapter 5

¹The Agricultural Project, *New Directions in Farm, Land and Food Policies* (Washington, D.C.: Conference on Alternative State and Local Policies, 1980).

Chapter 23

¹Permaculture® is a registered trademark of the Permaculture Institute and is registered as a trade service mark in Massachusetts by Dan Hemenway, New England Permaculture Consultancy.

²Although Holmgren and Mollison originally coined the term *Permaculture* by combining *permanent* and agriculture, Mollison describes the design philosophy as one of *permanent culture.* Permaculture® encompasses far more than agriculture, involving the location of roads, and, indeed, virtually any biological, geographical, physical and/or social consideration in the design or retrofit of a human settlement.

³A design philosophy such as Permaculture® would be unnecessary if the environment were not in serious need of repair. It is an approach to environmental healing.

⁴The Agricultural Project. *New Directions in Farm, Land and Food Policies.* Washington, D.C.: Conference on Alternative State and Local Policies, 1980.

Annotated Bibliography

———◆◆◆———

Berkshire Garden Center, Inc. *Gardening Pains?* Stockbridge, Mass. 1981. Has a unique brochure on adaptive garden tools for the gardening handicapped.

Britz, Richard et al. *Edible City Resource Manual.* Los Altos, Calif.: William Kaufman, 1980.

Brooklyn Botanic Garden. *Community Gardening: A Handbook.* Brooklyn, N.Y.: Brooklyn Botanic Garden, 1979.

"Children, Gardens . . . and Lead." *Mother Earth News* (July-August 1981).

Coe, Mary Lee. *Growing with Community Gardening.* Taftsville, Vt.: Countryman Press, 1978. One of the few resource books specifically for community gardeners, with a focus on community garden profiles from several localities.

Foster, Catherine Osgood. *The Organic Gardener.* New York: Vintage, 1972.

Gardens for All. 180 Flynn Ave., Dept. p45G, Burlington, Vt. Has a series of booklets on each family of vegetables; thorough and easy to follow for planting, tending, and harvesting information. Also publishes a bimonthly newsletter, *Gardens for All,* that contains gardening and organizational information relevant to community gardeners.

Institute for Local Self-Reliance. *Self-Reliance.* 1717 Eighteenth Street, N.W., Washington, D.C. 20009. A bimonthly newsletter covering projects across the country and information relevant to ecological local development in food, energy, and other basic resources.

Mollison, Bill. *Permaculture Two.* Tasmania, Australia Publications, 1979. TAGARI, P.O. Box 96, Stanley, Tasmania 7331, Australia.

Mollison, Bill and Holmgren, David. *Permaculture One.* Winters, Calif.: International Tree Crop Institute, U.S.A., 1981.

New Alchemy Institute. 237 Hatchville Road, East Falmouth, Mass. 02536. Publishes applied research in tree crops, organic agriculture, solar greenhouse; publishes *The Journal of the New Alchemists;* hosts workshops and tours on Saturdays in the summer.

Newcomb, Duane G. *Postage Stamp Garden Book: How to Grow All the Food You Can Eat in Very Little Space.* Los Angeles: J. P. Tarcher, 1975.

Stout, Ruth and Clemence, R. *The Ruth Stout No-Work Garden Book.* Emmaus, Pa.: Rodale Press, 1971.

175

Suffolk County Extension Service. *Lead in the Soil: A Gardener's Handbook.* Boston: University of Massachusetts, 1979.

Trust for Public Land. 254 W. 31 Street, New York, N.Y. 10001. Has several short handbooks useful for researching, acquiring, and developing land as a community land trust. These include *Land Inventory Handbook, Neighborhood Real Estate Primer, Neighborhood Land Trust Handbook,* and *Citizens Action Manual.*

People Power. U.S. Office of Consumer Affairs. Washington, D.C.: 1979. A resource manual of ideas, organizations, and programs across the United States that promote self-help, community-based services in their localities.

Yepson, Roger B. *The Organic Way to Plant Protection.* Emmaus, Pa.: Rodale Press, 1976.

Index

acrylics, solar use of, 124
Agricultural Organization Society, 13
airborne lead pollution, 51
Almeida, Ray, 8, 10, 158
annuals, planting, 89
aphids, 98

bank property, 52
barrel-and-hose irrigation, 79
biological pest management, 156
Boston Urban Gardeners (BUG), 158–62
bylaws, corporate, 133

chain link fence, 84
chicken- or turkey-wire fence, 82
children, activities for, 117–20
    older teens, 120
    six- to fourteen-year-olds, 118–19
    tots, 118
coalition building, 136–39
community canning center, 144–46
community gardens
    cost estimates, 163–64
    experiences of, 158–62
    history of, 11–15
    introduction to, 3–10
    local food systems, 127–62
    meaning of, 4–5
    organizing, 19–25
    public role, 32–41
    reason for, 6–10
    resources, 26–31
    site selection and development, 43–94
    tending, 95–125
community land trust, 131
Community Self-Reliance, Inc., 45–46
compost, 69–73, 107
    creating, 70–71
    decomposition, 71–72
    purpose of, 69–70
    spreading over the garden, 72–73
    See also soil
conservation
    energy, 37
    public role and, 36
    of water, 75–76
conservation holdings, 135
containments, testing for, 62
copper chromium arsenate, 122
copper naphthenate (Cuprinol), 122
corporate land ownership, 130–31

DeVaughn, Booker, 161
Dorchester Gardenlands Preserve (DGP), 161
Dupont, Bud, 97
dwarf trees, 155

energy conservation, 37
Environmental Protection Agency (EPA), 136

farmers' markets, 40–43

Federal Community Food and Nutrition
     Program, 146
fences, 48, 81–87
   assistance from contractors, 84–85
   basic design, 83
   maintenance and repairs, 85–87
   plantings along side of, 85
   types of, 82–84
fertility, soil, 68
fertilizers, 93–94, 107
   list of, 171
fiberglass reinforced polyester, 24
food co-ops, 147–49
Frost, Robert, 81–87
fruits, perennial, 155–56
funding, sources of, 30–31, 167

garden coordinators, 108–12
   continuing education and information,
      110–12
   organization, 108–10
   physical resources, 110
garden plot application, 25
gathering spot, 55–57
germination, 105
glass, solar use of, 122–24
Gomez-Ibanez, Nina, 117–20
Grady, Patricia, 81–87, 129–35
Graves, Isaac, 32, 125
"guinea gardens" (England), 12

Harding, Lloyd A., 4
herbaceous perennial plants, 153–54
herbicides, 99
hoses, 77

initial site preparation, 53–54
intercropping, 101

jug irrigation system, 78–79

Kahn, Charlotte, 11–15, 45–52, 53–58
Kincaid, Gareth, 3, 113

land, locating, 45–52
   evaluating a site, 47–51
   exploring options, 46
   securing a site, 51–52
land ownership alternatives, 129–35
   community land trust, 131
   conservation holdings, 135
   corporate bylaws, 133
   corporate ownership, 130–31
   long-term leases, 134–35
   nonprofit corporation, 131–33
   tax-exempt status, 133–34
landscape planting, 88–94
   with annuals, 89
   knowing plant's form, size, and growth
      rate, 89–90
   maintenance, 93–94
   and microclimates, 91
   with perennials, 89
   plant material and planting, 92–93
   root systems, 91
   seasonal features, 90
   site preparation, 91–92
layout, 53–58
   gathering spot, 55–57
   initial site preparation, 53–54
   limitations and "specialization," 57–58
   paths, 54
   plot sizes, 57
leaching, defined, 8
lead contamination of soil, 63–66
leases, long-term, 134–35
Libby, Pat, 158–62
Libby, Tom, 5, 45
Liberty Gardens, 14
lime, 72
Lueders, Jane, 88–94

maintenance
   fence, 85–87
   plant, 93–94
manure, 105, 107
Martin, Martha, 108–12, 113–16

Massachusetts Board of Education, 13–14
microclimates, 91, 151
Morse, Jean E., 53–58, 74–80
mulching, 75–76, 93, 106, 153
multifunctional plants, 152

Naimark, Susan, 81–87, 97–107, 121–25, 147–49
Narragansett Indians, 15
natural features, 50
nonprofit corporation, how to form, 131–33
nutrient cycling, 153–54
nutrient tests, 61–62

organic matter. *See* compost
organizing, basic steps in, 19–25
    coordination and responsibility, 23–25
    finding people, 20–21
    first meetings, 22

paths, 54
perennials
    fruits and vegetables, 155–56
    herbaceous, 153–54
    permaculture, 151
    planting, 89
permaculture, 150–57
    biological pest management, 156
    in a community garden, 157
    microclimate modification, 151
    multifunctional plants, 152
    nutrient cycling, 153–54
    perennial fruits and vegetables, 155–56
    perennials, 151
    soil improvement, 152–53
    tiered planting, 154
    tree crops, 155
pesticides, 99, 106
pests and diseases, 98–99, 106–7
pH test, 61
plants, 97–107
    along street side of fence, 85

deciding what to grow, 99–100
    preparations, 103–6
    seeds and seedlings, 102–3
    tips on planting, 100–102
Plexiglas, solar use of, 124
plot sizes, 57
polyethylene, solar use of, 124
preparation, site, 91–92
Project Greenbean, 146
proposal writing, 168–70
property
    privately owned, 52
    publicly owned, 52
pruning, 94
public role, 32–41
    agriculture, 36
    beautification and conservation, 36
    charting a course, 38–41
    community development, 35
    economic development, 37
    education, 36
    energy conservation, 37
    nutrition, 36
    political arena and, 37–38
    recreation and therapy, 36
    stalking the government, 33–37

rain gauge, 77
raised beds, 100–101
Redlich, Susan, 32–41, 136–39
resources, 26–31
    human, 27–28
    networking, 28–29
    physical, 110
    recycling, 165–67
    scrounging, 29–30
    sources of funding, 30–31, 167
root system, 91
Rosenmiller, David, 150–57
Roth, Linda, 59–68, 69–73

scatter-seeded planting, 105
Scharfenberg, Virginia, 45–52

securing a site, 51–52
seeds and seedlings, 102–3
sheet mulching, 153
site, slope of, 48
site selection and development, 43–94
    compost, 69–73
    fencing, 81–87
    landscape planting, 88–94
    layout, 53–58
    locating land, 45–52
    soil, 59–68
    water, 74–80
soil, 59–68
    assessing, 59–60
    buying new, 66–67
    evaluating condition of, 47
    fertility, 68
    improvement of, 152–53
    lead contamination, 63–66
    testing, 60–62
    See also compost
soil supplements, 72
solar devices, 121–25
Southwest Corridor Community Farm
    (BUG), 161
spacing plants, 101
spring seedlings, 105
squash vine borer, 99
succession planting, 101
sunlight, 49–50

tax-exempt status, applying for, 133–34
tending the garden, 95–125
    keeping it going, 108–12
    with kids, 117–20
    plants, 97–107
    solar devices for, 121–25
    vandalism, 113–16
thinning plants, 105
tiered planting, 154
tilling wet soil, 72
tools, 103

topsoil, how to calculate needs, 67
transplanting, 105
tree crops, 155
trees and shrubs, best seasons for
    planting, 92
Tsongas, Paul, 32, 125

U.S. General Services Administration,
    39
U.S. War Gardens, 14

vacant lots, how to evaluate, 47–51
    access to water, 50–51
    distance from major streets, 51
    existing vegetation and natural
        features, 50
    fencing, 48
    slope, 48
    soil condition, 47
    sunlight, 49–50
    visibility, 51
vandalism, 113–16
vegetables, 155–56
    nutritious kinds to grow, 172
vertical growing, 101–2

Wagner, Judith, 3–10, 19–25, 113–16
water, 74–80
    access to, 50–51
    conservation, 75–76
    making the most of, 80
    systems of watering, 76–80
    weight per gallon, 78
Watkins, Francis, 5
Watson, Greg, 26–31, 40–43
weeds, 152
wire mesh fence, 82
Women in Agriculture, Food Policy and
    Land Use Reform, Inc., 144–45
wooden fences, 82–84
World War I War Gardens, 14
World War II Victory Gardens, 14–15